SECOND EDITION

How Do You Know?

Using • Math • To • Make • Decisions

 KENDALL/HUNT PUBLISHING COMPANY
4050 Westmark Drive   Dubuque, Iowa 52002

Edited By
Holly Hirst
James Smith

This book has been printed directly for the author's camera ready copy.

CONTENTS

PREFACE

Intent

This book is designed to be used in a general education course for liberal arts majors. We chose material that actively engages students, focusing on four applications that show the broad spectrum of areas that use mathematical understanding in some way. Three of these applications were chosen because of their applicability to students' own lives, and one because of it's pervasive use in areas as diverse as agriculture, industry and government.

Approach

The approach used in this text is to present problems and try to solve them. The answers to the problems are rarely "3.25", but more often "the best course of action is . . . because . . ." On the way to these solutions, students will learn some algebra and trigonometry, but there is no intensive review of hundreds of algebra factorizations and simplifications. Applications requiring algebra and trigonometry are used as central themes, and the math skills needed to analyze the problems are reviewed when needed. The computer or calculator is used whenever possible to minimize the amount of computation and maximize the amount of intuitive understanding.

In Chapter One, trigonometric functions are viewed as tools for everyday measurement rather than as beautiful periodic curves. In Chapter Two, mathematical finance formulas are considered in the context of personal finances and investments. In Chapter Three, statistics is introduced to help make students educated consumers of data. In Chapter Four, linear inequalities and optimization are discussed to show how prevalent mathematics is in industry.

Acknowledgments

We wish to thank the Mathematical Sciences Department for unflagging support. We owe many people a special thank you for their tireless work. Principle authors of the text and web material:

Jay Bennett	Holly Hirst	Jimmy Smith
Nancy Sexton	Hutch Sprunt	

Other contributors and class testers:

Theresa Early	Marty Gambrell
Gary Kader	Meg Moss
Heather Starr	Todd Tipton

Students who provided inspiration for the title and the cover illustration:

Andrew Foltz	Warren Potts	Tori Terrizzi

We also thank the 3000 students who have taken the course using our text, contributing suggestions and comments. Their ideas were very helpful in shaping the latest version of this text.

Holly Hirst and Jimmy Smith
May 1998

INTRODUCTION ▪▪▪▪▪▪▪▪▪▪▪▪▪▪▪▪▪▪▪▪▪▪▪▪▪

For most of you this is your first and last college mathematics course. This will be different from your previous mathematics courses in a number of ways:

- You won't memorize many formulas. Neither will you be required to do 47 repetitive problems each night. There won't be nearly as many problems for you to do, but each of them will require considerable work and thought; answers will usually be paragraphs rather than numbers.

- Every topic will relate to real problems that you may encounter in your daily life.

- You will often work in groups, or at least pairs, rather than by yourself.

- You will have formal labs and write up detailed laboratory reports.

- You will use computers and calculators extensively.

- We will teach no algebra before its time; i.e. algebraic skills won't be taught except as they relate to solving a problem requiring those particular skills.

- We expect you to write in the book a lot. Doing the "you try it" problems is critical to your success in the course.

What then will help you be successful in this course? PARTICIPATE!! Students who passively sit and want to have material fed to them will not appreciate the course, nor will they be appreciated. You must try the problems, mess with the computer, write your answers in paragraphs, speak up in class with questions or answers, and generally be an **active learner**. This means you will work hard, which you expected since this is a mathematics course. But your efforts will not be spent in solving equation after equation and in manipulating complex algebraic expressions. Instead you will use mathematical principles to evaluate possible courses of action, to reach reasonable conclusions, and then communicate your results.

Standards in this course are high, but reasonable. "A" work will go beyond minimal responses to assignments; A work will be correct solutions to problems accompanied by well-written, grammatical English explanations. Routine responses that are basically correct with only an occasional error get a "C". Indolence will receive its appropriate reward. Careful thinking and hard work will be rewarded; there isn't some magic mathematical genetic predisposition that will carry you through if you have it or doom you if you don't.

We expect you to enjoy this course. That may be a bold statement for some of you who have struggled with mathematics in the past, but you can enjoy mathematics if you participate and give it a chance. There are many opportunities to be creative in your thinking and communication; taking advantage of these will deeply enrich your experience. We hope the material, the method of presentation, the cooperative effort with the other students, and the support and encouragement of the teacher will all work together with your own efforts to make this a constructive and positive finale to your mathematical career.

1   TRIGONOMETRY AND MEASUREMENT

How do you measure the height of a tall tree? How can you determine the height of a mountain or the depth of a valley? How can you measure the distance across a lake or a canyon? Measurements like these, virtually impossible with a tape measure, are often easy using techniques of **trigonometry**. We find these distances by imagining the unknown length to be one side of a triangle. In fact, the word trigonometry comes from the Greek words meaning "triangle measure." Applying our knowledge of trigonometry, we will discover that triangles are very helpful mathematical tools for real world measurement.

A general knowledge of geometry can come in handy in all sorts of situations. Many of the products people buy are purchased in quantities based on the size of the area to be covered or the perimeter to be encompassed. Buildings are brimming with irregularly shaped rooms, walls, and roofs needing careful measurement to be built, painted or shingled. We'll learn how a little knowledge of triangles and rectangles can help in these measurements and calculations as well.

The History of Trigonometry

As early as 500 BC, ancient Babylonians and Egyptians used trigonometry for measurement when building or laying out crop fields. But the most advanced and prevalent use of trigonometry was in the study of the stars. Greek astronomer Hipparchus, who is credited with giving us our 360° measure of a circle, documented systematic use of trigonometry in 200 BC. The Hindus and Muslims further developed the use of trigonometry for astronomy by 400 AD. The first systematic European treatment was by Johann Muller (1436-1476). Georg Joachim von Lauchen (1514-1576) was the first to define trigonometric functions as ratios of sides of a right triangle, the way we'll define them later.

Since that time trigonometry has had a wide application in many areas of mathematics and science. Disciplines as diverse as physics and wood working use trigonometry for making precise measurements. The trigonometric functions are being used today to further the frontiers of our understanding of mathematics and the world around us.

THE GEOMETRY OF POLYGONS

1

Before introducing the trigonometric functions, we review some basic definitions from geometry that we'll assume you have seen before.

- **Angle** - the intersection of two line segments
- **Parallel Lines** - lines that lie in the same plane and don't intersect
- **Polygon** - a closed plane figure bounded by line segments
- **Triangle** - a three-sided polygon
- **Right Triangle** - a triangle with a 90° (right) angle
- **Hypotenuse** - the longest side of a right triangle, opposite the right angle
- **Similar Triangles -** a set of triangles having the same sized angles
- **Rectangle** - a four-sided plane figure in which sides intersect at right angles
- **Area** - the space contained inside of a plane figure
- **Perimeter** - the distance around the edge of a plane figure
- **Cylinder** - a solid in which all cross sections in one direction are the same
- **Volume** - the space contained inside of a solid

Triangles play a key role in many of our measurement tasks. Let's look at triangles and their properties in more detail. Three triangles are shown below with their respective measurements. Each angle is labeled with a capital letter and the side opposite the angle is labeled with the lower case of the same letter. The right angle is denoted with a small square. If several angles meet at the same point, the notation $\angle ABC$ will be used to indicate the angle made by line segments AB and BC. These conventions will be followed throughout the chapter.

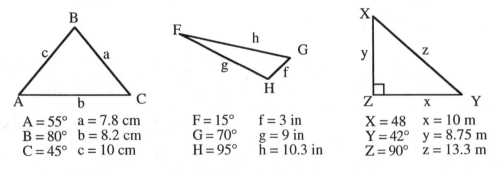

A = 55° a = 7.8 cm	F = 15° f = 3 in	X = 48 x = 10 m
B = 80° b = 8.2 cm	G = 70° g = 9 in	Y = 42° y = 8.75 m
C = 45° c = 10 cm	H = 95° h = 10.3 in	Z = 90° z = 13.3 m

In the example above the sides were measured using a variety of distance measurements: Centimeters, inches and meters. All of the angles were measured in degrees, °. Recall these common terms used to describe angles:

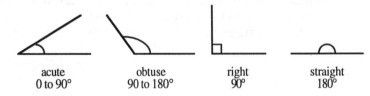

acute   obtuse   right   straight
0 to 90°   90 to 180°   90°   180°

We will need several properties concerning the sides and angles of triangles for our measurement exercises.

> ## The measurements of the three angles of a triangle add up to 180°.

This first property or **theorem** is particularly useful in finding one angle when we know the other two. An easy way to demonstrate this property is to:

1) Cut several triangles of differing shape and size out of paper.
2) Tear each triangle into three pieces so that there is one vertex on each piece.
3) Put the vertices together and let the sides of the original triangle touch.
4) Observe the straight angle formed when the angles are added together.

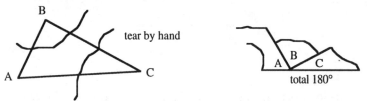

We refer to this property as a **theorem** because it is not a basic fact from geometry thus a **proof** must be given before we can assume it to be true. A mathematician would not call the paper ripping demonstration a proof, because it only shows the property for the particular triangle we cut out of the paper. A proof must not rely on properties of a specific triangle. Most proofs start with a given set of statements that are known to be true; each subsequent statement must follow logically from the ones that come before it, with the property in question as the concluding statement. Let's prove the property about the sum of the angles in a triangle using another useful geometric property as "given:"

> ## Parallel lines cut by a transversal have equal alternate interior angles.

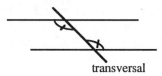

transversal

How does this help to prove that the sum of the angles in a triangle is 180°?

To prove the sum of the angles in a triangle is 180°, consider the following. Given: Parallel lines cut by a transversal have equal alternate interior angles.

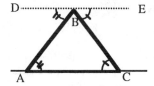 Consider the triangle ABC with the segment DE through the vertex B and parallel to side AC. Since DE is parallel to AC, the other two sides of the triangle, BC and BA, are transversals that cut a set of parallel lines.

From the parallel line property, $\angle DBA = A$ and $\angle EBC = C$.

Note that $\angle DBE = 180°$, since it's straight; thus

$$\angle DBA + \angle ABC + \angle EBC = 180°.$$

Substituting: $A + B + C = 180°.$ □

3

Notice that this proof works regardless of the shape of the triangle we draw. The parallel line can still be drawn and the parallel line property will still apply. In fact, the picture is really just given for clarity. The words contain all of the important parts of the proof. The □ is used to signal the end of the proof.

We used the parallel line property in this proof, so we'd better check to see if it is a true statement. How could we prove the parallel line property? We'll need to start with a "given" statement, in this case one of the basic **postulates** of plane geometry: *If parallel lines are cut by a transversal, corresponding angles are equal.*

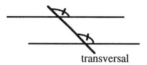

transversal

A postulate or **axiom** is a fundamental statement assumed to be true without proof. Some people use this postulate as the definition of parallel lines. Notice that this is almost the parallel line property we need. We'll draw a picture to get the main idea for the proof:

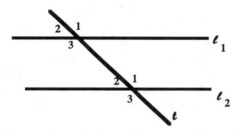

In this picture, we've labeled equal angles from the parallel postulate with the same numbers. Since ℓ_1 is straight, $\angle 1 + \angle 2 = 180°$, and since t is straight, $\angle 2 + \angle 3 = 180°$, so $180 - \angle 1 = 180 - \angle 3$, thus $\angle 1 = \angle 3$. This needs a little cleaning up but it is essentially the proof we were after. Can you rewrite this carefully in English, justifying each step? See Exercise 30 at the end of the section.

YOU TRY IT 1.1

Find the missing angles below.

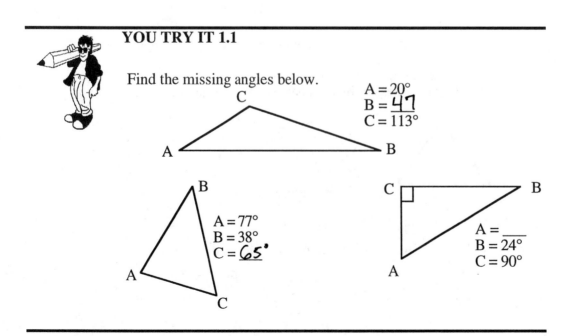

$A = 20°$
$B = \underline{47}$
$C = 113°$

$A = 77°$
$B = 38°$
$C = \underline{65°}$

$A = \underline{}$
$B = 24°$
$C = 90°$

Below is another extremely important theorem from geometry. We will use this one many times.

For any right triangle with hypotenuse labeled c: $a^2 + b^2 = c^2$.

This is referred to as the **Pythagorean Theorem**, in honor of Pythagoras (550 BC) who worked out the first known geometric proof of this property. We'll investigate a proof of this property at the end of the section. This property is useful for finding one side of a right triangle when we know the other two. The two shorter sides of a right triangle are often referred to as **legs**.

Example 1.1: What is the length of the hypotenuse c in this triangle?

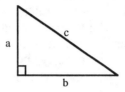

Solution:
$$3^2 + 4^2 = c^2$$
$$25 = c^2$$
$$5 = c$$

It doesn't matter what letters we use to label the sides. The critical feature of the equation in the Pythagorean relationship is that the hypotenuse is alone on one side of the equation. To remember this, use your common sense: If two sides (squared) are going to add up to (the square of) the third side, the **longest** side must be the one that is equal to the sum.

The triangle in Example 1.1 is special in that the arithmetic works out so nicely; this is not always the case.

Example 1.2: The length of the hypotenuse of a right triangle is 14 feet, and the length of one leg is 12 feet. How long is the other leg?

Solution: Using the Pythagorean Theorem again,
$$a^2 + 12^2 = 14^2$$
$$a^2 + 144 = 196$$
$$a^2 = 52$$
$$a = \sqrt{52} \approx 7.21 \text{ feet}$$

A note on the number of digits to report in your answer:

The exact answer to this problem is $\sqrt{52}$; the best approximate we can give (using a calculator) is 7.21110255. In practice it is not necessary to use all of these digits in the final answer; in fact, if our given measurements are only accurate to a few digits, using nine or ten digits in the answer is silly. We need to use common sense when reporting answers. Consider this situation: We measure two sides of a triangle with a meter stick on which the smallest marks are centimeters. At best we might be able to *estimate* these two sides to the nearest half-centimeter. Should we give a ten digit approximation for the third side after calculating it? Of course not! The nearest centimeter or half-centimeter is really the best we should hope for.

The moral of this story is: Your final answer should only have as many digits as the least accurate number in the calculation. So what should you do when working on problems? You may assume (unless specifically directed otherwise) that the numbers given in the problems are exact measurements. In this case, you may round to two decimal places in the last step to report your answer. You will practice both ways of thinking in the problem below.

YOU TRY IT 1.2
Find the length of the unknown side.
a) Assume that the numbers given are exact and round to two decimal places.
b) Assume that the numbers given are approximate measurements of the lengths and report the answer that makes sense.

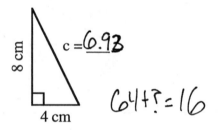

$c = \underline{6.93}$

$64 + ? = 16$

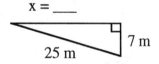

$x = \underline{\quad}$

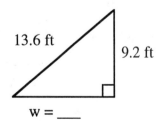

$w = \underline{\quad}$

YOU TRY IT 1.3
Solve the following triangles for all unknown parts.
Round your answer to 2 decimal places.

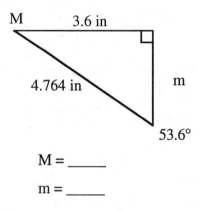

M = _____

m = _____

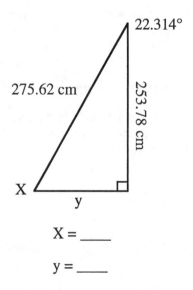

X = _____

y = _____

We are interested in how these rules apply to realistic situations. Let's consider an example.

Example 1.3: John is putting up a 100 foot radio antenna. The manufacturer recommends attaching guy wires 81 feet up the antenna and anchoring them in the ground 53 feet from the base of the antenna. How long will each guy wire have to be?

Solution: First sketch a picture and locate the triangle.

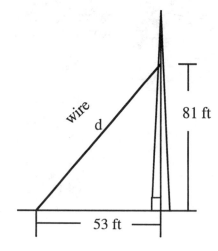

$$53^2 + 81^2 = d^2$$
$$9370 = d^2$$
$$96.80 = d$$

Each guy wire will need to be 96.80, or approximately 97, feet in length.

When making a mathematical model of a real world situation, we always have to make some assumptions. For instance, in the last example we assumed that the radio tower forms a right angle to level ground around it. We also assumed the guy wire is pulled so tight that it is straight. In any real world problem, we try to carefully consider our assumptions and interpret the results accordingly. In this problem John would want to buy a little extra wire; how much would you suggest?

YOU TRY IT 1.4

A surveyor wants to know how far it is across a pond. She measures off a right triangle as in the diagram below. What is the length of AC across the pond?

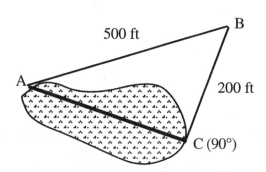

Sometimes finding the length of a side of a triangle is only the first step in a problem. Consider the following situation.

A home owner wants to plant grass seed to improve his lawn. The seeding rate given on the package is 1 pound per 1000 square feet, so he needs to figure the area of his lawn in order to know how much seed to buy. The lawn is rectangular with a small triangle attached as in the picture at right. Measurements are in feet.

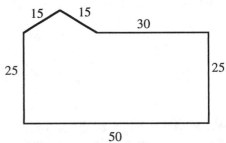

Where should we start?

The area of a rectangle is the length times the width.

The rectangle in the picture above has length 50 feet and width 25 feet, so the area of just the rectangle is $50 \times 25 = 1250$ square feet. How about the triangle part? We need a formula for the area of a triangle. We'll develop the formula for triangles from the formula for rectangles. Consider the picture below:

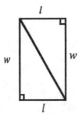

Notice that the rectangle has two right triangles in it, and the triangles are exactly the same size. So half the area of the rectangle has to be the area of one of the right triangles. This gives the area of a right triangle as one half of l times w, where l and w are the legs (or more commonly l is the base and w is the height). How about a triangle that is not a right triangle, as in our lawn problem? Consider the triangle below with an "altitude" drawn in perpendicular to the bottom or base of the triangle:

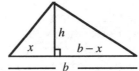

We've assumed that the altitude has height h and the base has length b, so the area of the entire triangle can be calculated as the sum of the areas of the two right triangles:

$$\frac{1}{2}xh + \frac{1}{2}(b - x)h = \frac{1}{2}xh + \frac{1}{2}bh - \frac{1}{2}xh = \frac{1}{2}bh$$

This should look familiar to you. Here is a general triangle formula:

The area of a triangle is one half the base times the height.

We have the formulas we need to finish off the homeowner's area calculation by finding the area of the triangular section of lawn. Drawing just the triangle, labeling the sides we know, and drawing in an altitude gives:

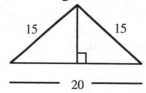

We need the height of the altitude. The altitude must cut the base in half, since the two other sides are the same length. (Can you show this using the Pythagorean Theorem? See Exercise 32.) The Pythagorean Theorem gives:

$$10^2 + h^2 = 15^2$$
$$h^2 = 125$$
$$h \approx 11.18 \text{ feet}$$

So the area of the triangular piece of land is $\frac{1}{2}(20)(11.18) \approx 112$ square feet.
Finally, we have the pieces needed for the total area of the lawn:

$$112 + 1250 = 1362 \text{ square feet.}$$

One last question: How much seed should the homeowner buy? Remember that the seeding rate was 1 pound per 1000 square feet. How many 1000s are there in 1362? $\frac{1362}{1000} = 1.362$. So to seed 1362 square feet he'll need 1.362 pounds of seed.

Example 1.4: The paint on the south side of a house has faded from exposure to the sun, and the homeowner wants to repaint. The brand of paint to be used covers 400 square feet per gallon. What is the area of the southern exposure as drawn, and how much paint should be purchased? Note: The windows are four feet by four feet and don't need paint!

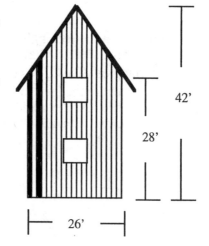

Solution: Ignoring the windows for a moment, we can split this figure into a 28 by 26 rectangle and a triangle with base 26 and height 42 - 28 = 14. Thus the area needing paint is:

$$(28)(46) + \frac{1}{2}(26)(14) - 2(4 \times 4) = 1438 \text{ square feet.}$$

The term subtracted is the combined area of the two windows, since the windows don't need paint. Notice that we didn't need the Pythagorean theorem for this triangle. Sometimes we will, and sometimes we won't.

To calculate the paint needed, we need to figure how many 400s there are in 1438. Dividing 1438 by 400 gives: 3.594 or 4 cans of paint . (Most stores don't sell fractions of cans of paint.)

YOU TRY IT 1.5

Dolores wants to estimate the cost of building her dream house. Her builder told her that his houses usually run $60.00 per square foot using exterior dimensions, with luxury appointments like stone fireplaces and hardwood floors costing extra.

a) Calculate the basic cost of the house given below.

b) If Hardwood floors cost an additional $10.00 per square foot, calculate the cost of the house if Dolores wants to use hardwood flooring in the family room, kitchen and breakfast nook.

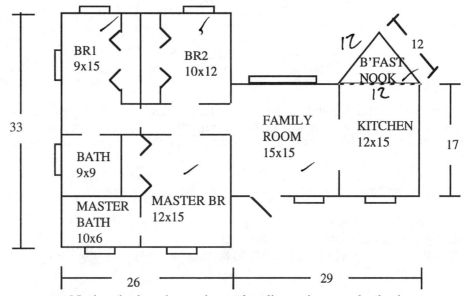

Notice the interior and exterior dimensions are both given.

Just as area calculations are needed for many measurement situations, so are volume calculations. The most common volume formula is that for a rectangular box: Volume = length x width x height. We can rewrite this as:

$$\text{Volume} = (\text{area of the base}) \times \text{height},$$

since the area of the base is length x width. This "formula" for calculating volumes can be generalized to any shape base provided that slices parallel to the base have the same shape as the base. Such solids are called cylinders. Here are some examples of cylinders with the base shape shaded. If we knew the area of the shaded bases and the heights we could multiply them to find the volumes of these cylinders – even the irregular cylinder!

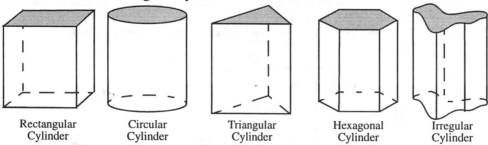

Rectangular Cylinder Circular Cylinder Triangular Cylinder Hexagonal Cylinder Irregular Cylinder

Notice the general naming convention: The base shape is the adjective. Of course, several of these have common names; rectangular cylinders are more often just called boxes, and the triangular and hexagonal cylinders (polygon bases) are called prisms.

Let's try a volume problem. Consider the following example.

Example 1.5: How much concrete should be purchased to pour a concrete driveway if the layout is given in the picture at the right and the thickness needs to be at least 8 inches?

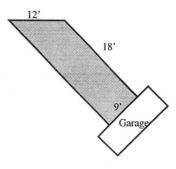

Solution: To answer this question, we can think of the driveway as a cylinder with base as given in the shaded part of the picture and height 8 inches = 0.6666667 feet. (It is extremely important to make sure our units are uniform -- all in feet or inches -- before proceeding.) This reduces the calculation to:

Amount of concrete needed in cubic feet = (area of driveway) x 0.6666667

To calculate the area, note that we appear to have a rectangle at the garage end of the driveway topped with a triangle at the street end. The rectangle is 9 by 18; so the area for that part of the driveway is 9 x 18 = 162 square feet. The triangle is a little more complicated:

We need to find h before we can use the triangle area formula. From the Pythagorean Theorem,

$$h^2 + 9^2 = 12^2$$
$$h = 7.937254.$$

Thus the area of the triangle is $\frac{1}{2}(9)(7.937254) = 35.71764$.

Now we're ready to finish the calculation:

$$\begin{aligned} \text{concrete needed} \quad &= (\text{area of driveway}) \times 0.6666667 \\ &= (162 + 35.71764) \times 0.6666667 \\ &= 131.81 \text{ cubic feet.} \end{aligned}$$

All of our examples so far have dealt with the lengths of sides of triangles or rectangles. We want to turn our attention to problems involving angles. We need some information about similar triangles before we can proceed.

Remember that two triangles having the same angles are similar. Below are two similar right triangles, ABC ~ DEF. Angles A = D and B = E.

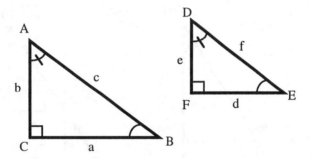

The sides of similar triangles are not necessarily equal, as is obvious in the picture above. However, the lengths of corresponding sides are related in the following way.

Corresponding sides of similar triangles are in proportion.

For the triangles above:

$$\frac{a}{d} = \frac{b}{e} \text{ or, rearranging, } \frac{a}{b} = \frac{d}{e}.$$

The second proportion implies that the ratio of the leg across from A (called the "side **opposite** A") over the leg next to A (called the "side **adjacent** to A") will be equal in similar right triangles, regardless of their sizes. Look at the three similar triangles below.

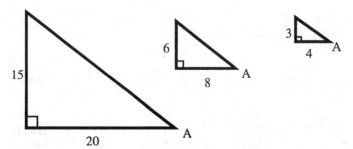

$$\frac{\text{opposite}}{\text{adjacent}} = \frac{3}{4} = \frac{6}{8} = \frac{15}{20}$$

If we used a protractor to carefully measure A, we would find it to be about 37°, (actually 36.87° to 2 decimal places). So any time we see a right triangle with 37° as one of its angles, we know that the relationship between the two legs can be written:

$$\frac{\text{side opposite A}}{\text{side adjacent A}} \approx \frac{3}{4}.$$

The size of the sides of the triangle makes no difference. You get a different ratio only if you start with a different angle!

Once again we have stated a property from geometry, so we ought to be sure it's true before going further. How should we start? Let's draw a picture and get some ideas. We'll assume that ABC is similar to DEF, where ∠A = ∠D and ∠B = ∠E, and ∠C = ∠F = 90°. Consider two arrangements of the similar triangles in a larger triangle:

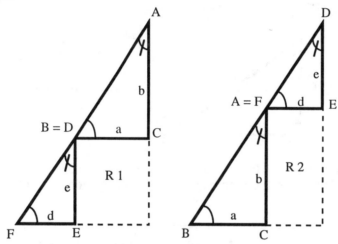

The large triangle formed in each case is the same. (Can you verify that these are always triangles and always the same size?) So the rectangles R 1 and R 2 must have the same area. What are their areas?

Area (R 1) = $a \times e$ Area (R 2) = $d \times b$.

So:

$$a \times e = d \times b$$

or

$$\frac{a}{b} = \frac{d}{e}.$$

This is exactly the proportion we need. We've left a few details to fill in, but this is essentially the proof we needed. Can you fill in the details and write up the proof in English? See Exercise 33.

Now that we have a proof, let's look at how this proportion of the sides can be used.

Example 1.6: Larry wants to know how high the top of his basketball backboard is. He backs up sixteen feet from underneath the backboard and measures the line-of-sight angle from the ground to the top of the backboard to be about 37°. How high is the top of the backboard?

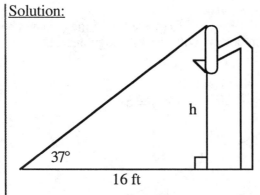

Solution:

Since the angle is 37°, we know that the ratio

$$\frac{opp}{adj} \approx \frac{3}{4}.$$

so we know that $\frac{h}{16} \approx \frac{3}{4}$

thus $h \approx 12$ feet.

What does the similarity property give us? In any two right triangles with the same acute angle, the ratios of opposite sides to adjacent sides are the same. We'll be able to use this idea to find the length of a side of a right triangle when we know one other side and an angle, rather than needing to know about two sides to get the third.

The ratio of opposite side length divided by adjacent side length is so useful that it has been given a special name: **Tangent**, often written

$$tan\, A = \frac{opp}{adj}.$$

This is our first trigonometric function! We'll investigate other useful side ratio relationships in the next section.

Exercise Set 1.1

In Exercises 1 through 4, solve the triangles for all unknown parts.

1)

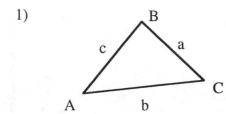

A = $55°$	a = 4.9 cm
B = 82°	b = 5.23 cm
C = 43°	c = 3.8 cm

2)

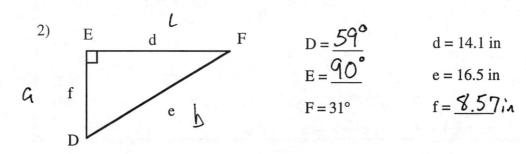

D = $59°$	d = 14.1 in
E = $90°$	e = 16.5 in
F = 31°	f = 8.57 in

Extra for Experts

The similar triangle properties can be used to prove the Pythagorean Theorem. Consider the right triangle ABC given below, with the line segment CD perpendicular to AB through angle C.

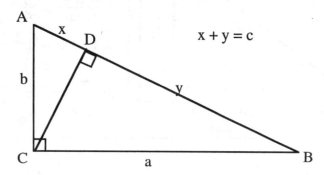

$$x + y = c$$

Claim: The two small triangles, ACD and CBD, are similar to the triangle ABC.

> Proof of the claim: We need to show that the angles in the small triangles match up with the angles in the large triangle. Since the three angles in any triangle add up to 180°, we only need to show that two of the angles match up.

Let's start by showing that triangles ACD and ABC are similar:

triangle ACD		triangle ABC
A	=	A
right angle at D	=	right angle at C

Now look at triangles CBD and ABC. They're similar too:

triangle CBD		triangle ABC
B	=	B
right angle at D	=	right angle at C

Thus: ABC ~ ACD and ABC ~ CBD, and our claim is proved.

Now we can mark the corresponding sides and write down the relationships:

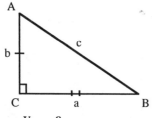 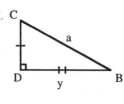 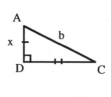

We have $\dfrac{y}{a} = \dfrac{a}{c}$ from CBD ~ ABC and $\dfrac{x}{b} = \dfrac{b}{c}$ from ACD ~ ABC.

Multiplication gives: $\quad x = \dfrac{b^2}{c} \quad$ and $\quad y = \dfrac{a^2}{c}$

Since $c = x + y$: $\quad c = \dfrac{b^2}{c} + \dfrac{a^2}{c}$

Multiply through by c: $\quad c^2 = a^2 + b^2 \quad \square$

3)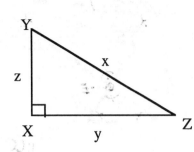

X = 90°   x = 19.95

Y = 72.1°   y = 18.98 m

Z = 17.9   z = 6.13 m

4)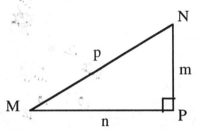

M = 48.2°   m = 60.37

N = 41.8°   n = 54 cm

P = 90°   p = 81 cm

In Exercises 5 through 14, draw a diagram (if not given) and solve the problem. Give your final answer accurate to two decimal places unless otherwise directed.

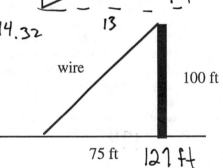

5) A draftsman is planning to use the template shown to the right. Using the given dimensions, find the length of the side marked X.

$13^2 + 6^2 = c^2$
$169 + 36 = c^2 = 14.32$

6) Lucy has to install a guy wire from the top of a 100 foot tower to an anchor 75 feet from the base of the tower. If she needs an extra foot on each end to secure the wire, how much wire should she buy?

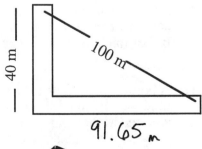

127 ft

7) Mark courteously walks around the corner on the sidewalk but John walks diagonally across the grass. If John travels 100 meters and Mark travels 40 meters on the first side, how much farther will Mark travel than John?

91.65 m

8) An architect needs to give an accurate measurement for the total amount of lumber (in feet) needed to build twenty trusses (wood is needed for all three sides of the triangular truss) with dimensions given in the figure.

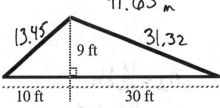

13.45   31.32

17

9) In order to calculate the distance across a pond, the surveyor has determined the dimensions in feet stated in the diagram on the right. Assume that AB is parallel to CD, and find the distance across the marsh without getting wet.

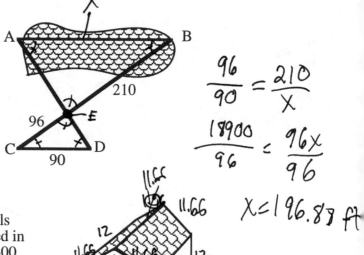

$$\frac{96}{90} = \frac{210}{X}$$

$$\frac{18900}{96} = \frac{96X}{96}$$

$$X = 196.87 \text{ ft}$$

10) A barn is being built according to the measurements in the picture. The exterior walls will be painted red and the roof will be shingled in black shingles. If one gallon of paint covers 400 square feet, how much paint is needed to put two coats on the walls of this barn? How many square feet of shingles are needed to cover the roof?

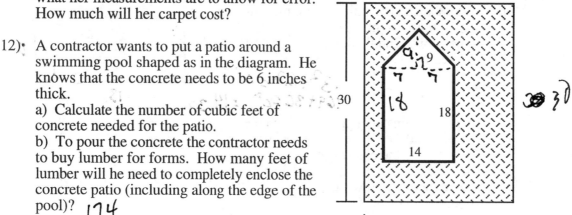

11) Dolores wants to carpet the bedrooms, family room and breakfast nook in her house (pictured in YTI 1.5 on page 11). The carpet she wants sells for $22.50 per square yard installed, and the carpet dealer told her to order an additional 3% over what her measurements are to allow for error. How much will her carpet cost?

12) A contractor wants to put a patio around a swimming pool shaped as in the diagram. He knows that the concrete needs to be 6 inches thick.
a) Calculate the number of cubic feet of concrete needed for the patio.
b) To pour the concrete the contractor needs to buy lumber for forms. How many feet of lumber will he need to completely enclose the concrete patio (including along the edge of the pool)? 174

13) Suppose the contractor from Exercise 12 needs to estimate the amount of concrete needed to pour the bottom of the pool. If the bottom must be 4 inches thick, how many cubic feet of concrete are needed?

14) A homeowner needs to calculate the cost of removing a hill of dirt with cross-section as pictured to the right. Each truck load of 50 cubic feet costs $20.00 to haul away. If the hill is 22 feet long with roughly the same cross-section, how much will it cost to haul?

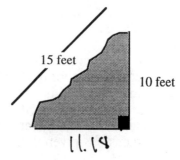

15 feet

10 feet

11.18

$$A_{\triangle} = \frac{1}{2} bh$$

$$= \frac{1}{2} b(10)$$

$$= \frac{1}{2} 11.18(10)$$

$$= 55.90 \times 22 = 1229.8$$

$$\frac{1229.8}{50} = 24.596 = 25 \text{ loads} = \$500.00$$

18

15) A surveyor needs to check his measurements for a triangular piece of land. One way to check is to see if the three angle measurements fit the triangle properties. From his measurements he has calculated the three angles in the triangle to be 86.21°, 51.12° and 42.33°. Has an accurate measurement of the triangle been made? *yes   179.66 = 180°*

16) Triangle properties can be helpful when dealing with other polygons. For example, we can answer the question "What is the sum of the angles of any pentagon (5-sided figure)?" Try this: Draw any pentagon and cut it into triangles so that the vertices of the triangles all coincide with the vertices of the pentagon. What is the answer? How about for an octagon (8-sided figure)?

A fundamental knowledge of basic triangle geometry is very helpful on the Graduate Record Examination (GRE), a test required for admission to many graduate schools. The questions below are similar in structure and content to those appearing in the quantitative section of the GRE. The directions for problems 17 through 27 are: Each of the following problems contains two quantities labeled A and B. Your task is to determine the relationship between the sizes of the quantities. Answer (A) if A > B, (B) if A < B, (C) if A = B, or (D) if the relationship between A and B cannot be determined from the information given. Make no additional assumptions; pictures are not drawn to scale.

17) Assume line ℓ is parallel to line *m.*

      A: x°   B: y°

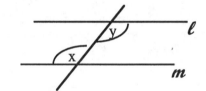

18) Given any triangle:

      A: The sum of the angles in the triangle   B: 90°

19) Given Δ ADC to the right.

      A: length of *AB*   B: Length of *BC*

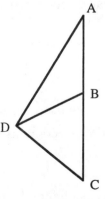

20) Given Δ ADC to the right.

      A: ∠ADB   B: ∠ADC

21) Given Δ ADC to the right.

      A: ∠ABD + ∠DBC   B: 180°

Figure for 19) through 22).

22) Given Δ ADC to the right.

      A: perimeter of Δ ADC   B: perimeter of Δ BCD

23) Given the triangle to the right.

A: $c^2 - b^2$ B: a^2

24) Given the triangle to the right with $a = 3$ and $b = 4$

A: c B: 10

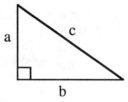

Triangle for 23) and 24)

25) $\triangle$ RST is similar to $\triangle$ OPQ in the picture to the right.

A: $OQ \times RS$ B: $RT \times OP$

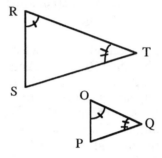

26) The figure at the right represents the floor of a certain room.

A: Area of the floor   B: 600 sq ft

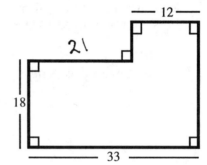

27) A 16 foot ladder leaning against a vertical wall with the base of the ladder 8 feet from the wall slips down the wall 2 feet, causing the ladder to slide away from the wall by x feet.

A: x   B: 2

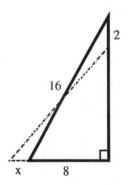

The GRE also contains multiple choice questions similar to 28 through 31 below.

28) According to the figure to the right, traveling directly from point A to point B, rather than through point C would save approximately how many miles?
a) 13   b) 4   c) 3   d) 2   e) 1

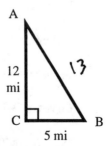

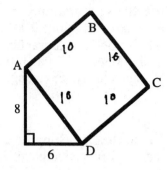

29) In the figure to the right, what is the area of the square ABCD?
a) 100   b) 64   c) 49   d) 36   e) 25

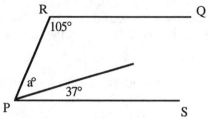

30) If RQ is parallel to PS then a =
a) 53   b) 68   c) 58   d) 38   e) 105

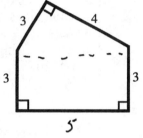

31) What is the perimeter of the pentagon?
a) 13   b) 21   c) 18   d) 23   e) 15

32) Write up a proof of the statement: Parallel lines cut by a transversal have equal alternate interior angles. (See page 4.)

33) Write up a proof of the statement: The area of a triangle is one half the base times the height. (See page 9.)

34) Write a proof of the statement: The altitude of an isosceles triangle bisects the base. (See page 10.)

35) Write a proof of the statement: Corresponding sides of similar triangles are in proportion. (See page 14.)

2

THE TRIGONOMETRIC FUNCTIONS

We saw in the last section that the **tangent** of an acute angle in a right triangle is the fraction formed by placing the opposite side over the adjacent side.

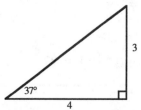

$$tan\,37° \approx \frac{3}{4} = 0.75$$

Your calculator already has the values of tangent and other trigonometric functions programmed in. Be sure your calculator is in **degree** mode (see the manual or ask your instructor if you're not sure), then press **37 TAN** (or **TAN 37** if you have a graphing calculator) and the display will read 0.7535541, which is about 3/4. We can use this proportion to help answer questions about right triangles in which an angle measurement is given.

Example 2.1: Suppose a ladder is leaning against a house and the base of the ladder forms an angle of 72° with the level ground. The bottom of the ladder is 10 feet from the building. How far up the building will the top of the ladder touch?

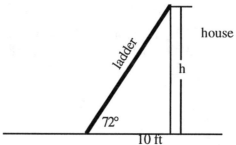

Solution: Since we have a right triangle and the question involves an acute angle and its opposite and adjacent sides, we can use the tangent to set up an equation. using our calculators (in degree mode) we find **72 TAN** is 3.0776835.

$$tan\,72° = \frac{h}{10}$$

$$3.07768354 = \frac{h}{10}$$

$$h = 30.7768354 \approx 30.78$$

The ladder will rest against the house almost 31 feet up. (Incidentally, how long is the ladder? It is a little over 32 feet.)

Example 2.2: Suppose the ladder is moved so that the top of the ladder is only 28 feet up on the side of the house and the angle of the bottom of the ladder with the ground is 65°. How far from the base of the house is the bottom of the ladder?

Solution: $tan\,65° = \dfrac{28}{d}$

so:

$$2.144507 = \frac{28}{d}$$

$$2.144507d = 28$$

$$d = 13.05661 \text{ or about 13 feet away}$$

YOU TRY IT 2.1
Suppose that a tree casts a shadow that is 40 feet long when the sun's rays make an angle of 67° with the ground. How tall is the tree? (Be sure to draw a picture.)

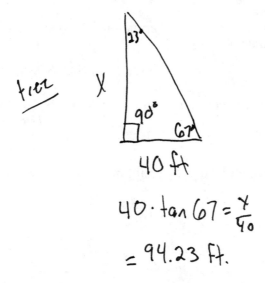

tree   X

40 ft

$40 \cdot \tan 67 = \frac{Y}{40}$

$= 94.23 \text{ ft.}$

There are other valuable relationships like this. The ones we will use are given below for an acute angle A in a right triangle:

tangent A $= \dfrac{\text{side opposite}}{\text{side adjacent}}$	**OR**	$\tan A = \dfrac{\text{opp}}{\text{adj}}$
sine A $= \dfrac{\text{side opposite}}{\text{hypotenuse}}$	**OR**	$\sin A = \dfrac{\text{opp}}{\text{hyp}}$
cosine A $= \dfrac{\text{side adjacent}}{\text{hypotenuse}}$	**OR**	$\cos A = \dfrac{\text{adj}}{\text{hyp}}$

In our approach to problems, we will use the relationship that helps us the most. Let's consider some examples.

Example 2.3: A = 28° and a = 3.5. Find side c.

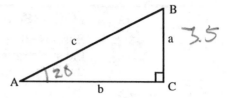

Solution: Here are several
alternative solution strategies.

1) It **is** possible to use our tangent equation to find side c, but it
involves two separate equations:

    Use $\tan 28° = \dfrac{3.5}{b}$ and solve for b, to find $b = 6.58254263$.

Use the Pythagorean Theorem to find c: $3.5^2 + 6.58254263^2 = c^2$

                                                   $7.46 \; = \; c$

2) The sine relation will yield the result in one step. $\sin A = \dfrac{opp}{hyp}$:

$$\sin 28° = \frac{3.5}{c}$$

$$c = \frac{3.5}{\sin 28°} \approx 7.46$$

.5317094317

3) Note that the cosine relation is not useful because there are 2
unknowns!

$$\cos 28° = \frac{b}{c}.$$

.766044443

Example 2.4: Sue is putting up guy wires to a tower. The manufacturer
recommends that the angle with the ground be 65°. If Sue plans to anchor the wires
40 feet from the base of the tower, how long will each wire
need to be?

.422618261

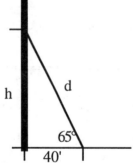

Solution: Let's consider the solution alternatives.

1) Suppose we jumped in and started using the
tangent function. This would allow us to find h and
then we could use the Pythagorean theorem to find
d.

2) Sine is worse; There are two unknowns,
opp $= h$ and adj $= d$.

3) What about cosine? $\cos 65° = \dfrac{40}{d}$, which gives $d = 95$ feet.

Example 2.5:

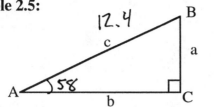

A = 58°
c = 12.4
b = _____

Solution: Let's try to decide in advance which formula would be most efficient. We have the angle and the hypotenuse, and we want to find the adjacent. The only formula with all those quantities in it is cosine's.

$$\cos 58° = \frac{b}{12.4}$$

$$12.4 \times \cos 58° = b$$

$$b = 6.57 \text{ (rounded.)}$$

Example 2.6: Find sine, cosine and tangent for both angle S and angle T below. Can you clearly identify the opposite and adjacent sides for each acute angle?

Solution:

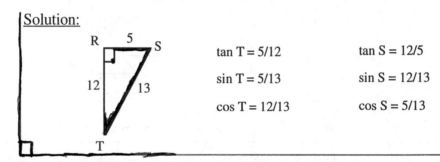

tan T = 5/12   tan S = 12/5

sin T = 5/13   sin S = 12/13

cos T = 12/13   cos S = 5/13

As we can see from the last four examples, drawing the picture first, and carefully labeling the angle and sides of interest, will help to determine which formula to use.

YOU TRY IT 2.2

Let's look at a couple of triangles and practice this last concept. Find the sine, cosine and tangent for each acute angle below.

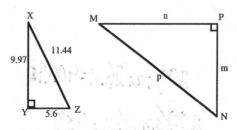

25

tan X = _____ tan Z = _____ tan M = _____ tan N = _____

sin X = _____ sin Z = _____ sin M = _____ sin N = _____

cos X = _____ cos Z = _____ cos M = _____ cos N = _____

Here are some tips for setting up triangle problems.

Choosing the Right Trigonometric Formula

1) We use the equation that involves two known parts of the triangle and the part whose size we are trying to find.

2) Remembering that the sum of the angles of a triangle is always 180°, we can find B given A in a right triangle: $A + B + 90° = 180°$
$$B = 90° - A.$$

3) If we are considering a right triangle from **the other acute angle**, the labeling changes. Look at the difference below:

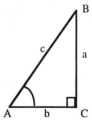

viewing from A
opp = a   adj = b
hyp = c (always)

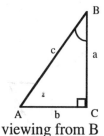

viewing from B
opp = b   adj = a
hyp = c (always)

YOU TRY IT 2.3

To measure the distance across a river from Point R to Point S, a surveyor measures 200 feet along the river bank on a line that forms a right angle with the line RS (see diagram). From Point T, she sights an angle of 63.8° between Point R and Point S. Find the distance across the river from Point R to Point S. Also find the distance ST.

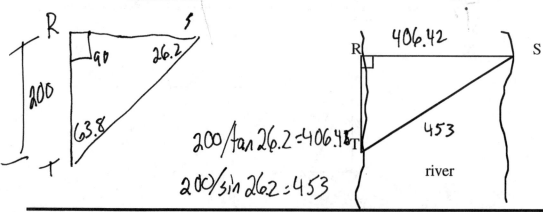

YOU TRY IT 2.4

According to the manufacturer, the maximum safe angle for a fireman's extension ladder to make with the ground is 70°. If the longest ladder on a particular truck is 50 feet, how high will that ladder reach above ground if used safely? (Draw the diagram, label the parts and solve.)

Finding an Angle

Sometimes the side information is given and the angle is unknown. To solve for the angle, we'll have to move, or "undo," the trigonometric function. This means that we need an inverse function for each trig function. Luckily our calculators have these functions built in. Let's look at some examples.

Example 2.7: Sarah's ladder reaches 22 feet up the side of her house when the bottom is 10 feet from the wall. What angle does the bottom of the ladder make with level ground?

Solution: $tan A = \dfrac{22}{10} = 2.2$

What now? Instead of wanting to find the tangent of an angle, we want to find an angle whose tangent is 2.2.

This is the INVERSE procedure from the previous problems. We need to "undo" the tangent. Fortunately, your calculator will find an INVERSE TAN, although the key strokes will vary between models. On most calculators, we find the angle by pressing

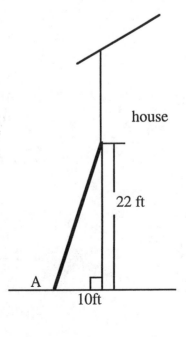

house

22 ft

A

10ft

2.2 INV TAN

To get 65.6°. The **INV** key simply accesses the commands written above the keys on the calculator. If you don't have the **INV** key, look for a **2nd**, **2ndF**, **Shift** or **ARC** key.

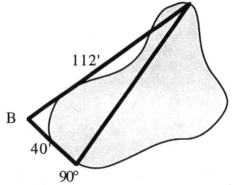

Example 2.8: Ralph is checking a survey map and finds the layout to the right. Putting his transit on the spot marked B in the diagram, what angle should he find if the map is correct?

Solution:

$$\cos B = \frac{adj}{hyp} = \frac{40}{112}$$

$$\cos B = 0.357142857 \quad \text{(use inverse cos!)}$$

$$B = inverse\ \cos 0.357142857 \approx 69.08° \text{ (rounded)}$$

YOU TRY IT 2.5

Find the angle to the nearest tenth of a degree for each:

if $\tan A = 0.79$ then A = _____

if $\tan B = 8.15$ then B = _____

if $\sin M = 0.87$ then M = _____

if $\cos R = 0.13$ then R = _____

YOU TRY IT 2.6

A 32 foot ladder is leaning against a vertical tree. If the bottom of the ladder is 10 feet from the tree, what angle does the bottom of the ladder form with level ground (to the nearest tenth of a degree)?

YOU TRY IT 2.7

A rental truck is parked on level ground. The truck bed is 4 feet above the ground. An 18-foot aluminum ramp is hooked to the back of the truck and used to carry furniture up into the truck. What angle does the bottom of the ramp make with the level ground (to the nearest tenth of a degree)?

In You Try It 2.7, the angle the ramp makes with the ground is called the **angle of elevation** of the ramp. The angle of elevation is always measured from a horizontal line <u>up</u>. In contrast, an **angle of depression** is always measured from a horizontal line <u>down</u>.

Example 2.9: Ralph is standing 120 feet from a statue. His transit is 3.5 feet high. The angle of elevation to the top of the statue is 26.6°. Find the height of the statue.

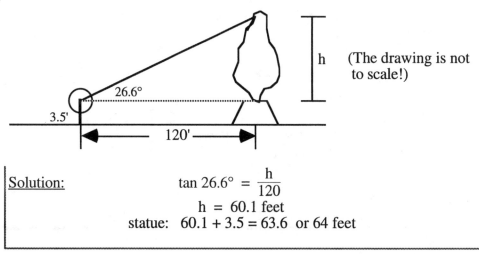

(The drawing is not to scale!)

Solution:
$$\tan 26.6° = \frac{h}{120}$$
$$h = 60.1 \text{ feet}$$
statue: $60.1 + 3.5 = 63.6$ or 64 feet

Example 2.10: Sighting out his sixth floor window, John finds the angle of depression to his car in the parking lot to be 39°. If his window is 64 feet above level ground, how far is his car from his residence hall?

Solution:

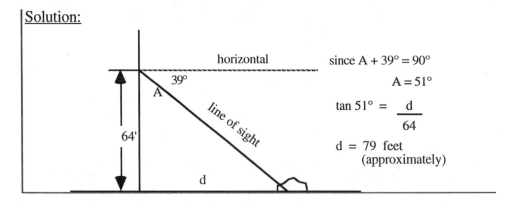

since $A + 39° = 90°$

$A = 51°$

$$\tan 51° = \frac{d}{64}$$

$d = 79$ feet
(approximately)

Correct placement of the angle of depression is very important. You can draw a horizontal reference line in your picture as we did in Example 2.9, or you can remember that angle of depression down from A to B is the same as the angle of elevation up from B to A. Why? Look at the picture below and think of the "parallel lines cut by a transversal" property.

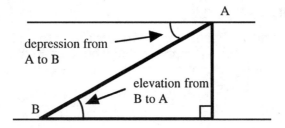

Example 2.11: From the top of a lighthouse on the shore, the angle of depression to a life raft on the ocean is 39°. If the top of the light house is 100 feet above sea level, how far off shore is the life boat?

Solution:

$$tan\, 39° = \frac{100}{d}$$

$$d\, tan\, 39° = 100$$

$$d = 123.49$$

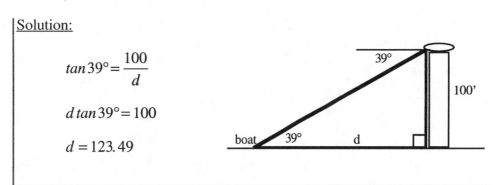

YOU TRY IT 2.8

From a boat on the water, the angle of elevation to the top of a lighthouse is 18.3°. If the top of the lighthouse is 150 feet above sea level, how far is the boat from the base of the lighthouse?

YOU TRY IT 2.9

The top of a lighthouse is 150 feet above sea level. The attendant finds the angle of depression to a boat out in the water to be 18.3°. How far is the boat from the base of the lighthouse?

Exercise Set 1.2

In Exercises 1 through 4, solve the triangles for all unknown parts.

1)

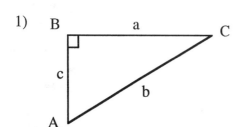

$A = \underline{62.96°}$ $a = \underline{754.44 \text{ ft}}$

$B = \underline{90°}$ $b = 847 \text{ ft}$

$C = \underline{27.04°}$ $c = 385 \text{ ft}$

7.43

2)

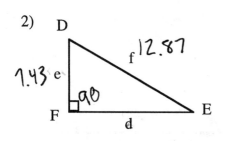

$D = \underline{ 54.74}$ $d = 10.508"$

$E = \underline{ 35.26}$ $e = 7.43"$

$F = \underline{90°}$ $f = \underline{12.87}$ 12.87

3)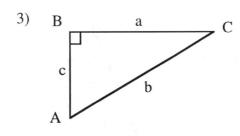

A = ___
B = 90°
C = ___

a = 102 ft
b = 398.28
c = 385 ft

10404   148225

4)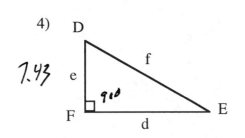

7.43

D = 80°
E = 10°
F = 90

d = ___
e = 7.43"
f = ___

In Exercises 5 through 28, draw a diagram (if not given) and solve the problem. Give your answer accurate to two decimal places unless otherwise directed.

5) Using a 4 foot tall transit, Sarah measures the angle to the top of a tree and finds it to be 37°. If she is 100 feet from the tree, how tall is the tree?

6) Tom leans his ladder up against a building so that the base of the ladder forms a 60° angle with the ground. If the top of the ladder is 20 feet up the building, how far from the wall is the bottom of the ladder?

7) Jane has a 32 foot ladder. If she leans it against a building so that the base forms an angle of 70°, how high up the building will the top of the ladder be?

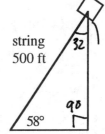

8) Sue is sitting in a field flying a kite high in the air. She has let out 500 feet of string. If the angle between the ground and her string in 58°, how high above the ground is the kite?

9) A guy wire is attached to the top of a radio antenna and to a point on horizontal ground that is 40 meters from the base of the antenna. If the wire makes an angle of 62.3° with the ground, approximate the length of the wire.

10) To measure the height of the cloud cover, a meteorology student shines a spotlight vertically up from the ground to the clouds. Using a transit from 1000 meters away, he measures the angle from level ground to the spotlight beam on the clouds and finds it to be 59°. Approximate the height of the cloud cover.

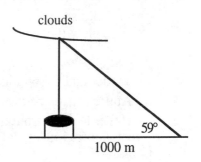

11) A rocket is fired at sea level and climbs at a constant angle of 75° through a distance of 10,000 feet. Approximate the rocket's altitude to the nearest foot.

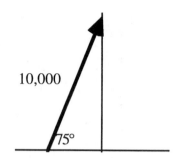

12) To find the distance d between two points P and Q on opposite shores of a lake, a surveyor locates a point R that is 50 meters from P such that RP is perpendicular to PQ as shown. Next, using a transit, the surveyor measures ∠PRQ as 72.6°. Find d.

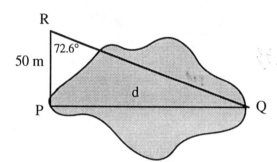

13) Mary stands directly across a river from a large tree. She walks along the river for 50 meters at a right angle to the line between where she stood (point A) and the tree. Using a transit she finds the angle from her path to the tree to be 57.4°. Find the distance from point A to the tree.

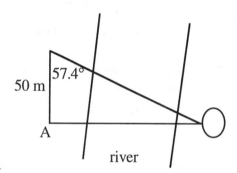

14) If the bases on a baseball diamond form a square that is 90 feet on a side, how far is it from home plate to second base?

15) A girl flying a kite holds the string 4 feet above ground level. The string of the kite is taut and makes an angle of 60° with the horizontal. Approximate the height of the kite above level ground if 400 feet of string is paid out.

16) Jane sights a billboard from 200 feet away. The angle of elevation to the top of the billboard is 14°, while the angle of elevation to the bottom of the billboard is 8.5°. How tall is the billboard itself (to the nearest tenth of a foot)?

17) A guy wire from the top of a 100 foot tower is 160 feet long. Find the angle the wire makes with the ground.

18) Sarah wants to build a ramp up to her porch. The porch is 4 feet above level ground and she wants the ramp to start on the ground 26 feet from the porch. What angle will the ramp make with the ground (to the nearest degree)?

19) A ladder 20 feet long leans against the side of a building. The angle between the ladder and the building is 22°.

a) Approximate the distance from the bottom of the ladder to the building.

b) If the distance from the bottom of the ladder to the building is increased by 3 feet, approximately how far does the top of the ladder move down the building?

20) A builder wishes to construct a ramp 24 feet long that rises to a height of 5 feet above level ground. Approximate the angle that the ramp should make with the ground.

21) The angle of elevation from a point P on the ground to an aircraft flying at an altitude of 5000 feet measures 22°. How far is it (in feet) from P to the point on the ground directly beneath the aircraft?

22) From a point 15 meters above level ground a surveyor measures the angle of depression to an object on the ground as 68°. Approximate the distance from the object to the point on the ground directly beneath the surveyor.

23) A swimming pool is 20 meters long and 12 meters wide. The bottom of the pool is slanted so that the water depth is 1.3 meters at the shallow end and 4 meters at the deep end. Find the angle that the bottom of the pool makes with horizontal at the shallow end.

24) What is the angle of depression from the top of a 170-foot lighthouse to a ship sailing 0.5 miles from the base of the lighthouse?

25) A vertical tower casts a shadow that is 300 feet long. At the same time, the shadow of a 20 foot tall building is 16 feet long.

a) Find the height of the tower.

b) Find the angle of elevation to the sun at the moment described, to the nearest hundredth degree.

3

MULTIPLE TRIANGLES

Understanding the principles of right triangle trigonometry enables us to handle a variety of interesting real world situations. Often the solution to a mathematical model may involve several interlinked triangles.

Example 3.1: Debbie and Joe want to measure the height of a building. They know the distance to the building on level ground. Their transit (a surveying instrument that measures angles) measures the angle to the top of the building to be 52°. Their transit is 4 feet above the ground.

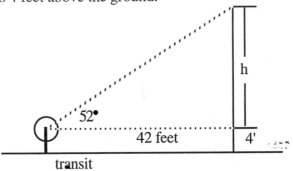

Solution: We see that the triangle formed by the transit head and the building does not have a side which is the total height of the building. We can solve, however, for the side opposite the 52° and simply add the extra 4 feet.

$$tan\,52° = \frac{h}{42}$$

h = 53.76 or about 54 feet (round to nearest whole), so the actual building height is 54 + 4 = 58 feet.

Example 3.2: Let's help Debbie and Joe with another one. First consider the situation:

building

We want to find the height of the building. What should we do? Consider placement of the transit, what can and can't be measured, how to get the most accurate measurement, etc.

Solution: Consider the adapted drawing and labels below. Joe and Debbie decided to do it this way, although other ways will also work.

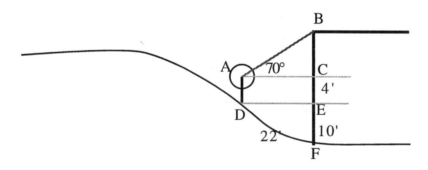

FIND DE: Locate a point C that you can reach from the ground and locate the transit horizontally across from C. Measure CF and subtract 4 feet to find EF. Now measure DF and use the Pythagorean Theorem to find DE.

FIND BC: Since DE = AC, we know AC. We read angle BAC from the transit and solve for BC, using the tangent function.

FIND HEIGHT OF BUILDING: We simply add
BC + CE + EF =
53.9 + 4 + 10 = 67.9 or about 68 feet

Joe and Debbie attacked the problem this way because they saw that CF was possible to measure. If it had been too high, they would have had to adjust their plan. How?

Example 3.3: On top of a building that is 95 feet tall there is a flagpole. How can we find the height of the flagpole without leaving the ground? (Before looking at the solution below, think through the situation and decide how you might do the problem using a transit.)

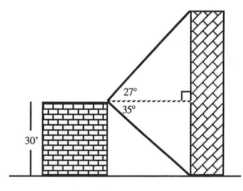

Solution: Here's one possible solution:

$$\tan 66° = \frac{h+91}{50}$$

$$50 \times \tan 66° = h+91$$

$$h \approx 21 \text{ feet}$$

There are a variety of ways to solve these problems. The way that makes sense to you and uses valid and reasonable steps is the best one to use. All problem solvers go through the basic steps of reading to understand the problem, brainstorming about solving using the tools available, trial and error efforts (checking for appropriateness and accuracy), and deciding on an answer. The methods used by two problem solvers may be different, but their conclusions will be the same.

Example 3.4: To estimate the height of a tall building next to his, Ted measured the angle of elevation from his roof to the top of the building and the angle of depression from his roof to the bottom of the building. Find the height of the taller building from the measurements given in the picture.

Solution: The first thing to do is decide on a plan of attack. If we find the opposite side for the 27° angle, and then add 30 feet to it, we should have the answer. We can't do this directly, because we don't know the adjacent side or the hypotenuse in the 27° triangle. Can we find one of those? Yes! We can use the 35° triangle to find the adjacent for the 27°; look at the new picture.

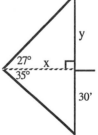

$$\tan 35 = \frac{30}{x} \text{ implies that } x \approx 42.8444.$$

Now, $\tan 27 = \dfrac{y}{42.8444}$, so $y \approx 21.83$ feet.

The height of the taller building next to Ted is about $30 + 21.83 = 51.83$ feet.

YOU TRY IT 3.1

Using his 3.5 foot transit, John stands 100.0 feet from a building that has an antenna on its roof. From the transit, he finds the angle of elevation to the top of the antenna to be 58.9° and the angle of elevation to the top of the building itself to be 53.4°. Find the height of the antenna.

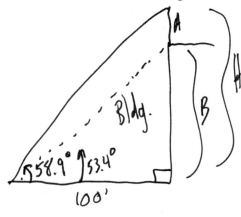

$$\tan 53.4° = B/100' = 134.65'$$

$$\tan 58.9 = H/100 = 165.77'$$

$$165.77 - 134.65 = 31.12' = \text{A}$$

Example 3.5: Debbie's and Joe's final task is measuring the height of a building without being able to get close to the building. (It must have a moat or something around it.) You may find a better way, but here's what Joe and Debbie decided to do. They found the angle of elevation to the top of the building using a 4 foot tall transit. Then they moved 30 feet closer and measured the new angle of elevation (see diagram). What then? Can you figure out what they planned to do next? Stop a minute and try it.

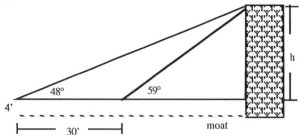

Solution: Here's one way to do it. **Read carefully and look over each step.** First, label the main points in the diagram and naming the distance from the second sighting to the building with the letter "x".

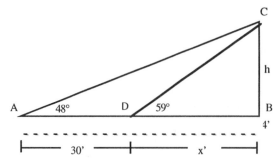

Consider the two main triangles formed at each sighting and labeled as such.

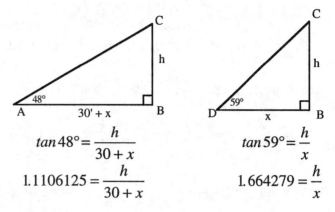

$$tan\,48° = \frac{h}{30 + x}$$

$$1.1106125 = \frac{h}{30 + x}$$

$$tan\,59° = \frac{h}{x}$$

$$1.664279 = \frac{h}{x}$$

Solving the second equation for h (multiplying by x), $1.664279x = h$.

Substitute $1.664279x$ for h in the left equation and solve for x:

$$1.1106125 = \frac{1.664279x}{30 + x}$$

$$1.1106125(30 + x) = 1.664279x$$

$$33.318375 + 1.1106125x = 1.664279x$$

$$33.318375 = 1.664279x - 1.1106125x$$

$$33.318375 = 0.5536665x$$

$$x = 60.177697$$

Use this value for x and back substitute in the original equation above on the right:

$$1.664279x = h$$

$$h = 100.15$$

CONCLUSION: $100.15 + 4$ (transit!) $= 104.15 \approx 104$ feet high

This problem required that we solve two equations in two unknowns by substituting one into the other. We can avoid having to deal with two unknowns by using a little geometry knowledge. Consider the picture drawn below for 3.5:

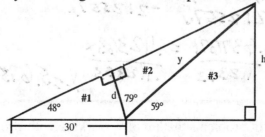

Notice the perpendicular labeled d; this divides the large triangle into three smaller triangles (numbered in the picture). The 79° angle can be found by noticing that the

other angle in triangle #1 must be 42°, and that 79 = 180 - 42 - 59. Here's the plan of attack to find h:

1) Find d using triangle #1.
2) Find y using triangle #2 with d from step 1.
3) Find h using triangle #3 with y from step 2.

1) From triangle #1: $sin\,48 = \dfrac{d}{30}$, so $d = 22.294345$.

2) From triangle #2: $cos\,79 = \dfrac{d}{y} = \dfrac{22.294345}{y}$, so $y = 116.84116$.

3) From triangle #3: $sin\,59 = \dfrac{h}{y} = \dfrac{h}{116.84116}$, so $h = 100.15$, same as before!

Either of these methods works well on two triangle problems with this type of picture. Here's one to try.

YOU TRY IT 3.2

You are directly approaching a volcanic island in the Pacific Ocean. You measure the angle of elevation to the highest point on the island to be 12°. After sailing 3 miles closer to the island, you measure the angle of elevation to be 18°. How high above sea level is the top of the island to the nearest hundred feet? (Recall that 1 mile = 5280 ft.)

$tan\,12° = \dfrac{H}{3+x}$ $(3+x)\cdot tan\,12° = H$

$tan\,18° = \dfrac{H}{x}$ $x \cdot tan\,18° = H$

$(3+x)\cdot tan\,12° = x\cdot tan\,18°$

$(3+x)\, .212557 = x\cdot 0.324920$

$.637671 + .212557x = .324920x$
$\quad\quad -212557v \quad\quad -212557x$

$\dfrac{.637671}{.112363} = \dfrac{.112363x}{.112363}$ $x = 5.6751$

$tan\,18° = \dfrac{H}{5.6751}$

$= 1.84\ mile.$

$1.84 \cdot (5280)$

$9700'$

There are many types of problems that involve working with two triangles and two unknowns. Here's another to try in which the triangles are back-to-back instead of nested. To start, label everything and write down any trigonometric relationships you see. You Try It 3.4 is yet another variation on two triangles.

YOU TRY IT 3.3

Bill and Sue are on opposite sides of a 300 foot wide canyon looking down at a monument in the bottom as in the picture below.
a) How far is each person from the base of the monument?
b) Assuming that the monument is in the deepest part of the canyon, how deep is the canyon?

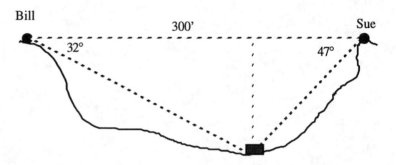

YOU TRY IT 3.4

From a 100 foot tall lighthouse on shore, an observer watches a boat sailing directly towards the base of the lighthouse. If the angle of depression from the observer to the boat changes from 31° to 48° in five minutes, how far has the boat traveled?

There are many times when a little knowledge of trigonometry can help you find an unmeasurable height or distance. Now you have this knowledge in your tool belt!

Exercise Set 1.3

In Exercises 1 through 32, draw a diagram (if not given) and solve the problem. Give your answer accurate to two decimal places unless otherwise directed.

1) A TV antenna is located on top of a 20-foot-tall garage. From a point on the ground that is 90 feet from the base of the garage, the antenna subtends an angle of 12° as shown. Approximate the length of the antenna.

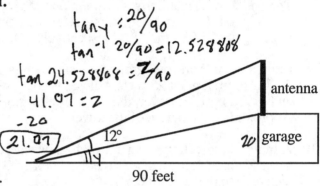

2) From a point A that is 8.20 meters above ground level, the angle of elevation to the top of a building is 31.33°, and the angle of depression to the base of the building is 12.84°. Approximate the height of the building to the nearest tenth of a meter.

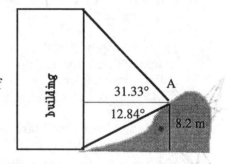

3) A motorist traveling along a level highway observes that the angle of elevation to the top of a mountain is 10°. After driving another 10 miles directly toward the mountain, the angle of elevation changes to 70°. Approximate the height of the mountain above level ground to the nearest foot.

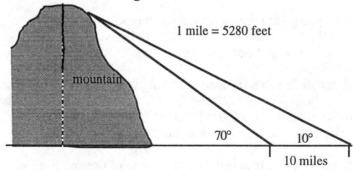

4) In order to find the height of a tree that is across the river from him, John measures the angle of elevation to the top of the tree from two places. First he backs up from the river and measures the angle of elevation to be 32.4°. Then he moves 50 feet closer to the tree and finds the angle of elevation to be 37.8°. How tall is the tree?

43

5) A rocket is fired at sea level and climbs at a constant angle of 85° for a distance of 13,000 feet. Approximate its altitude to the nearest foot.

6) A device for measuring cloud height at night consists of a vertical beam of light, which makes a spot on the clouds. The spot is then viewed from a point 135 feet away from the device. The angle of elevation is 67°. Find the height of the clouds.

7) A local fire safety regulation states that the maximum angle of elevation for a rescue ladder is 72°. If the fire department's longest ladder is 110 feet, what is the maximum safe rescue height under this regulation?

8) The angles of elevation from a point on the ground 175 feet from the base of a building to the top and bottom of a television antenna located on the top of the building are 71.07° and 68.3° respectively. How tall (in feet) is the antenna?

9) From a point P on level ground the angle of elevation to the top of a tower is 26.7°. From a point 25 meters closer to the tower and on the same line with P and the base of the tower, the angle of elevation to the top is 53.5°. Approximate the height of the tower.

10) A motorist, traveling along a level highway at a speed of 60 km/hr directly toward a mountain, observes that between 1:00 pm and 1:10 pm the angle of elevation to the top of the mountain changes from 10° to 70°. Approximate the height of the mountain above the road (to the nearest hundredth of a kilometer).

11) From the top of a building that is 220.0 feet high, an observer looks down on a street. Two cars are parked on the street, one behind the other, in line with the observer. If the lines of sight of the observer to the cars are 28.25° and 36.35° below horizontal, what is the distance between the two cars?

12) From the top of a cliff 600 feet above the plain, a sentry observed a stagecoach whose angle of depression was 10°. In line with the stagecoach he saw some bandits whose angle of depression was 7°. How far from the coach are the bandits? If the stagecoach is going 1250 feet per minute and the bandits 1670 feet per minute, will the bandits catch the stagecoach before it reaches the safety of the base of the cliff? What is the time difference (either way)?

13) Two men standing 200 feet apart have a flagpole between them. The angle of elevation to the top of the pole for one of them is 25°, for the other it is 30°. How tall is the pole and how far is each man from the base of the pole?

14) There is a brush fire in a field near a straight highway. Two observers on the highway are 3 miles apart with the fire off to one side between them. One sights the fire at an angle of 20° from the highway, and the other at 30°. How far is the fire from the highway and from each observer?

15) From the top of a 400 foot building the angle of depression to the top of a second building is 43°. If the buildings are 220 feet apart, how tall is the second building?

16) A rocket is fired vertically into the air. Three seconds after firing, an observer 1.2 miles away measures its angle of elevation at 6.5°. 8 seconds later the angle is 52°. What is the distance traveled by the rocket in the 8 seconds?

17) The seating floor of a theater is inclined 12° from the horizontal under-floor. If the dimensions of the under-floor are 140 feet (deep) by 85 feet (wide), find the number of square feet of carpet needed to cover the seating floor.

18) An architect is designing a ramp for a stadium that is to be built so that it reaches a height of 15 feet over a 100 foot run. Braces will be placed every 25 feet as in the diagram. Find the height of the three interior braces.

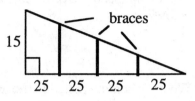

19) Heights make Billy nervous, but he gets on the 100 foot diameter Ferris wheel anyway. After he gets on, the wheel goes around through an angle of 57°. How high is Billy?

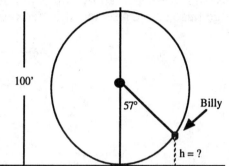

20) A carnival operator wants his roller coaster to have a climb of 150 feet in altitude before the exciting descent begins. He doesn't want the slope of the track to be more than 42°. What is the minimum length of that part of the roller coaster track? How much horizontal length must the operator have to set his track on, assuming that it is straight?

CIRCLES

Often when called upon to measure or build a structure, we must deal with curved edges. For example, architects and designers find curves an appealing change from lines, planes and angles in buildings. Such curved edges are often arcs of circles, so knowing a little circle geometry can help us handle these. What is the definition of a circle? The set of all points in a plane that are a fixed distance from a single point is called a circle. We call the fixed distance the radius and the fixed point the center. We get the familiar picture from this definition, with center labeled O and radius R.

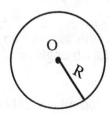

The formulas for the area and circumference (the perimeter) of a circle are:

The area of a circle is equal to πR^2, and the circumference is $2\pi R$.

What happens if we only have a piece of a circle to deal with, like the sliver pictured below?

We are often interested in the area of the sliver or in the length of the circular arc. These arcs are specified by number of degrees of the central angle, $\angle AOB$ in the wedge (also called a sector) pictured below. (O is the center of the circle.)

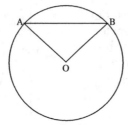

The length of the curve can be obtained from the size of the central angle and the radius of the circle using a proportion along with the following fact from geometry.

The total arc measure of a circle is 360°.

With this fact we can form a proportion relating measuring a circle using angles to measuring a circle using circumference:

$$\frac{\text{arc length}}{\text{circumference}} = \frac{\text{arc degrees}}{360°}$$

So what about the area? Well, we can use a proportion like the one above to find the area of the wedge given by $\angle AOB$, and then we can subtract off the area of the triangle. The proportion we need is:

$$\frac{\text{sector area}}{\text{circle area}} = \frac{\text{arc degrees}}{360°}$$

Now, how big is the triangle? Look again at the picture:

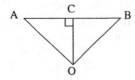

We have two equal right triangles with a known angle and side (OA and OB are radii of the circle). We use our trusty triangle area formulas and then we'll be finished.

Before looking at how these formulas will help with making measurements, let's review π, that interesting number appearing in the area and circumference formulas.

π (pronounced "pie" as in banana cream and usually spelled "pi") can be defined in many ways. It was probably first discovered as the ratio of the circumference of the circle to its diameter (the distance all the way across the circle through the center, which is 2R).

$$\pi = \frac{\text{circumference}}{\text{diameter}}$$

In the last two sections, we looked at interesting ratios that didn't change when the size of the triangle changed, such as tangent, sine and cosine. π is a ratio that doesn't change when the size of a circle changes. How might we investigate this? Try the following experiment. We'll need some string, a ruler, and a compass to draw some circles.

1. Draw several circles, each with a different diameter, or use the circles below.
2. Lay a piece string carefully around the edge of the circle and mark the distance around the circle. The more careful you are the more accurate your results will be.
3. Straighten the string out and measure it with a ruler.
4. Divide your circumference measurement by the diameter. What numbers did you get? They should be very close -- just a bit bigger than 3.

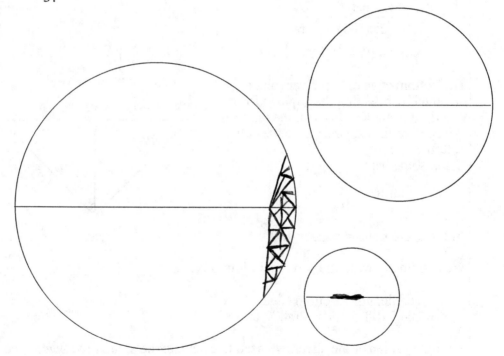

Pi has been the subject of much study by mathematicians. Many approximations have been proposed. The first few below should look familiar.

$$3.14 \qquad \frac{22}{7} \qquad 3\frac{1}{8} \qquad 3.14159265358979323846\ldots$$

Notice that these numbers all agree to at least one decimal place. $3\frac{1}{8} = 3.125$ (an estimate from 2000 BC Babylon) and $\frac{22}{7} = 3.\overline{142857}$. The last approximation in the list above is the exact value of π to 24 decimal places (25 digits). We will rely on our calculator's approximation, which should be accurate to about 10 digits.

So what is the value of π exactly? Mathematicians have debated this since ancient times, and finally proved in 1766 AD that it is a non-terminating, non-repeating decimal number. Hence we cannot ever write down all of the digits of π. Such numbers are called *irrational* numbers. π is one of the most famous; $\sqrt{2}$ is another. Let's try some measurements involving circles.

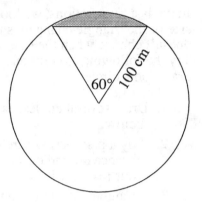

Example 4.1: Find the perimeter and area of the shaded portion of the circle.

Solution: To find the perimeter we need the arc length and the length of the bottom edge of the shaded figure. Using the arc length formula above, we find

$$\frac{\text{arc length}}{\text{circumference}} = \frac{\text{arc degrees}}{360°}$$

$$\frac{\text{arc length}}{2\pi(100)} = \frac{60°}{360°}$$

$$\text{arc length} = 104.7198 \text{ cm.}$$

The bottom edge of the sliver can be obtained by bisecting the angle to form two right triangles. Look at the picture. We need to find x, since 2x is the entire bottom edge.
Using some trig:

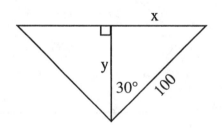

$$\sin 30° = \frac{x}{100}, \text{ so } x=50,$$

and thus the bottom edge is 100.

We can finally write the perimeter of the sliver as

$$\text{arc length} + \text{bottom edge}$$
$$= 104.7198 + 100 = 204.72 \text{ cm}$$

To find the area of the sliver, we need to calculate the area of the sector (the entire 60° wedge) and then subtract the area of the triangle. Let's start with the sector:

$$\frac{\text{sector area}}{\text{circle area}} = \frac{\text{arc degrees}}{360°}$$

$$\frac{\text{sector area}}{\pi \times 100^2} = \frac{60°}{360°}$$

$$\text{sector area} = 5235.98776$$

To find the area of the triangle we need the base and the height. We found the base to be 100 cm (when we calculated the perimeter) so now we need the height, which is y in the picture. We'll use trig again!

$$\cos 30° = \frac{y}{100}, \text{ so } y=86.60254 \text{ cm.}$$

Now let's assemble the area of the shaded part:
= area of sector - area of triangle
= area of sector - 1/2 × base × height
= 5235.98776 - 0.5 × 100 × 86.60254
= 905.86 cm²

Here's a more complicated example to look at and a few to try.

Example 4.2: We have purchased a 62 inch diameter hot tub for which we need to pour a concrete foundation. We also decide to enhance the existing deck with a circular step as pictured. All of this will be done with concrete.

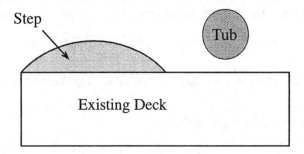

Here are the measurements:

- The concrete foundation for the hot tub should be 9 inches thick
- The step is formed from a 55° arc from a 15 foot diameter circle
- The step should be 6 inches thick.
- Concrete is sold in cubic yards, so we need to report the final answer that way.

Solution: We have two volumes to calculate: The tub foundation and the step. Let's do one at a time.

Foundation: This is a 9 inch thick cylinder with circular base radius 31 inches. Since the answer will need to be in yards eventually, let's change now:

$$31 \text{ in} \times \frac{1 \text{ yd}}{36 \text{ in}} = 0.8611111 \text{ yd and } 9 \text{ in} \times \frac{1 \text{ yd}}{36 \text{ in}} = 0.25 \text{ yd}$$

$$\text{volume} = (\text{area of base}) (\text{height}) = \pi(.8611111)^2(.25) = .58238 \text{ yd}^3$$

Bottom Step: This is a 6 inch thick cylinder with base shaped like a sliver from a 7.5 foot radius circle. The area of the entire 55° sector (sketched below) can be found from the proportion:

$$\frac{\text{sector area}}{\text{circle area}} = \frac{\text{arc degrees}}{360°}$$

$$\frac{\text{sector area}}{\pi(7.5)^2} = \frac{55°}{360°}$$

$$\text{sector area} = 26.99806 \text{ ft}^2$$

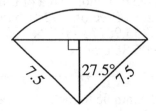

We have the area of the sector; next we need to find and subtract off the area of the triangle part to get the sliver only. One of the right triangles has hypotenuse 7.5 feet and angle 27.5°. We need the base (which is the side opposite our known angle) and the height (which is the side adjacent to our known angle). Back to trigonometry:

$$\sin 27.5° = \frac{\text{opp}}{7.5}, \text{ so opp} = 3.463115. \text{ The base is } 3.463115 \text{ ft.}$$

$$\cos 27.5° = \frac{\text{adj}}{7.5}, \text{ so adj} = 6.652581. \text{ The height is } 6.652581 \text{ ft.}$$

Now the area of the right triangle is

$$\tfrac{1}{2}\text{ base} \times \text{height} = \tfrac{1}{2}(3.463115) \times (6.652581) = 11.51933 \text{ ft}^2$$

Okay, we're ready to assemble our answer for the volume of the step:

$$
\begin{aligned}
\text{volume} \ &= \text{area of base} \times \text{height} \\
&= (\text{area of sector - 2} \times \text{area of right triangle}) \times \text{height} \\
&= (26.99806 \text{ - 2} \times 11.51933) \times 0.5 \\
&= 1.9797 \text{ ft}^3
\end{aligned}
$$

Oops! This is in feet and we'll need yards. Do we need to start over?? No! Let's change now:

$$1.9797 \text{ ft}^3 \times \frac{1 \text{ yd}}{3 \text{ ft}} \times \frac{1 \text{ yd}}{3 \text{ ft}} \times \frac{1 \text{ yd}}{3 \text{ ft}} = .073322 \text{ yd}^3$$

Notice that we had feet **cubed** so we had to do the unit change three times. We know we're done when the units we don't want have all canceled out.

Now to get the total amount of concrete needed, we add the foundation amount and the step amount to get:

$$0.58238 + .073322 = .655702 \text{ yd}^3$$

YOU TRY IT 4.1

Find the cost of building a circular concrete patio area around a circular pool as in the illustration below, if the diameter of the pool is 24 feet and the patio area is to be 6 feet wide and 8 inches deep. Concrete costs $30.00 a cubic yard delivered, poured and smoothed.

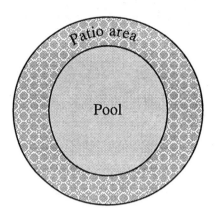

YOU TRY IT 4.2

Farmer Brown needs to estimate the amount of feed he has stored for winter. His silo, a 10 foot diameter circular cylinder, is filled two-thirds full with corn and his barn is filled to the rafters with hay bales. How many cubic feet of corn and bales of hay does he have, if a bale of hay is about 2 feet by 1.5 feet by 4 feet?

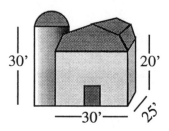

Occasionally, the surface area of a cylinder is needed. To find the surface area formula, we can think this way: Cut the cylinder open and lay it flat. We get a rectangle with dimensions height by perimeter of base:

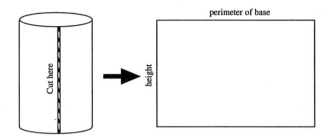

So if we need the surface area of a cylinder, we need to find the height of the cylinder and the perimeter of the base. If the base is a circle we would need the circumference.

Example 4.3: We have a circular turret room that we wish to paint. The diameter of the floor is 14 feet and the height of the walls is 10 feet. The paint we are going to use suggests two coats with a coverage of 400 square feet per one gallon can. How much paint do we need?

> Solution: We are painting the inside of a circular cylinder of radius 7 feet and height 10 feet. The surface area would be:
> $$\text{surface area} = \text{circumference} \times \text{height}$$
> $$= 2\,\pi\,R \times \text{height}$$
> $$= 2\,\pi\,(7) \times (10)$$
> $$= 438.82 \text{ square feet}$$
>
> We are applying 2 coats so we really need to cover 879.64 square feet. To find the number of gallons of paint, we multiply:
> $$879.64 \text{ ft}^2 \times \frac{1 \text{ can}}{400 \text{ ft}^2} = 2.2 \text{ gallons of paint.}$$
>
> If we can only buy whole cans of paint, we would need 3 cans.

Exercise Set 1.4

1)   Find the areas and perimeters of the shaded figures given below. Assume all circles have radius 1 foot.

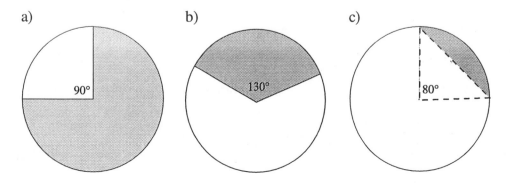

a)   b)   c)

90°   130°   80°

2) Find the areas and perimeters of the shaded figures given below. Assume all circles have radius 1 cm.

a)　　　　　　　　　b)　　　　　　　　c)

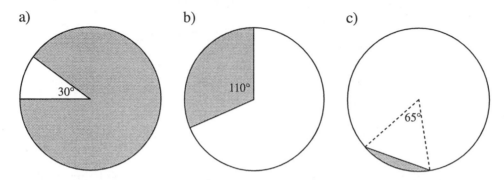

3) We want to build a circular paddock to hold 20 sheep. If each sheep needs about 10 square feet of space, what diameter circle would we need? How much fence would we need?

4) We have planned a garden and want to buy fencing. The garden area is shaped like a 10 by 12 foot rectangle with a half circle at each 10 foot end. How many feet of fence are needed to completely enclose the garden?

5) Farmer Brown's silo is a circular cylinder with base diameter 12 feet and height 27 feet. What is the capacity of this silo? If he can get $5.00 a bushel for corn and his silo is full of it, what is his revenue if one bushel of corn is about 2 cubic feet.

6) The Stewarts have just installed a new circular pool. It is 24 feet across and 4 feet deep. They want to fill it with clean water from the public water supply and have been told that they will have to pay 6 cents per gallon. If one gallon of water takes up about 0.16 cubic feet of space, how much will they pay to fill their pool?

7) We wish to build a circular silo with internal diameter 10 feet. How much concrete will we need to pour the foundation, if we only need a 1 foot wide and 1 foot deep ring on which the silo walls will sit? Assume the 4 inch thick silo wall rests on the middle of the ring as in the picture.

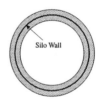

8) We wish to build a 10 foot wide patio around a circular swimming pool as in the picture. If the pool is 30 feet across and the patio needs to be 4 inches thick, how much concrete do we need?

9) We have a fenced in horse jumping ring of radius 20 feet, and we want to paint the 4 foot high solid fencing around the ring on both sides. How many square feet of fencing are there? If each can of premium exterior paint can cover 200 square feet, how many cans should we buy?

10) We are refurbishing our circular in-ground swimming pool and need to put a new coat of concrete sealer on the sides and bottom. If the pool is 30 feet across and a uniform 5 feet deep, how many square feet will we have to cover? If the sealer costs $22.50 per gallon and one gallon covers about 100 square feet, how much will it cost? (Assume that we can buy exactly the quantity we need.)

A fundamental knowledge of basic circle geometry is very helpful on the Graduate Record Examination (GRE), a test required for admission to many graduate schools. The questions below are similar in structure and content to those appearing in the quantitative section of the GRE. The directions for problems 11 through 14 are: Each of the following problems contains two quantities labeled A and B. Your task is to determine the relationship between the sizes of the quantities. Answer (A) if A > B, (B) if A < B, (C) if A = B, or (D) if the relationship between A and B cannot be determined from the information given. Make no additional assumptions; pictures are not drawn to scale.

11)   Assume that the area of the circle shown is 25π.

   A: x   B: 5

12)   Assume that the circumference of the circle shown is 25π.

   A: the radius   B: 5

Picture for
11 and 12

13)   Assume that the diameter of the circle is 2.

   A: area of inscribed   B: 4
        polygon ABCD

14)   Assume that the area of the circle is 20π.

   A: length of segment CD   B: 5

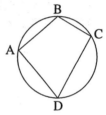

Picture for
13 and 14

The GRE also contains multiple choice questions similar to 15 through 18 below.

15)   If the area of the inscribed rectangle shown is 18, then the circumference of the circle is

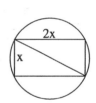

   a) $\sqrt{45}$ b) 6   c) 45π   d) $\pi\sqrt{45}$ e) 6π

16)   If the area of the circle is 5π, then x in the rectangle is

   a) 1   b) 2   c) 3   d) 4   e) 5

17)   If the circumference of a circle is between 10π and 20π, which of the following could be the area of the circle?

   a) 60   b) 210   c) 420   d) 660   e) 780

18)   If the area of a circle is between 10π and 20π, which of the following could be the circumference of the circle?

   a) 5   b) 15   c) 25   d) 35   e) 45

CHAPTER SUMMARY   5

One Last Thing: Tips for Handling Real World Problems

All of the measurement problems we have encountered thus far have been pretty well defined. Unfortunately, the real world doesn't work that way. We are usually confronted with a situation that we have to simplify or clarify before we can attack it. Consider the following question:

How much better is it to kick a football field goal from
the 10 yard line than the 30 yard line?

On the surface, there is no indication that we might answer this question using trigonometry. We need to pin down the question a little. What is "better"? Are we just looking at distance? If so the answer is easy, but most kickers can kick a ball far enough. What else is there? Let's draw a picture from the overhead perspective.

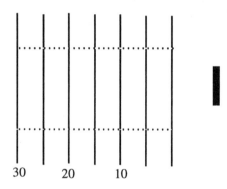

Clearly, it's going to matter where the player is on the 10 or 30 yard line. Let's assume that he's in the middle of the field. It's also going to matter how far apart the uprights on goal post are and how far past the end of the field they are. These things are standard for most college fields: Goal post uprights are 24 feet apart and 10 yards past the end of the field. Let's look at the picture again.

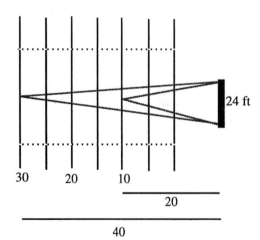

Now we start to see what happens. The angle of success for the kicker at 10 yards is larger than the angle of success at 30 yards! How much larger might make a good measure of "better". We've reduced the original question to a trigonometry problem. We can solve for the angles and report a solution. (See problem 10.)

As we have seen in this football example, once we carefully specify the problem we can use geometry and trigonometry to solve many kinds of measurement problems.

Key Ideas

General Triangles
- The measurements of the angles in a triangle add up to 180°.
- The area of a triangle is equal to one half the base times the height.
- Corresponding sides are in proportion for similar triangles.
- The altitude of an isosceles triangle cuts the base in half.

Right Triangles
- (Pythagorean Theorem) If the hypotenuse is labeled c and the legs are a and b, then $a^2 + b^2 = c^2$.
- The tangent of an angle is equal to the opposite side divided by the adjacent side. (tan A = opp/adj)
- The sine of an angle is equal to the equal to the opposite side divided by the hypotenuse. (sin A = opp/hyp)
- The cosine of an angle is equal to the adjacent side divided by the hypotenuse. (cos A = adj/hyp)

Angles of Elevation and Depression
- The angle of elevation is always measured from a horizontal line up.
- The angle of depression is always measured from a horizontal line down.
- The angle of depression from A down to B is equal to the angle of elevation from B up to A.

Circles and Arcs
- The circumference of a circle is 2π times the radius. (C = 2πR)
- The area of a circle is the radius squared times π. (A = πR^2)
- The length of a circular arc can be found using the ratio
  arc length : circumference = arc degrees : 360
- The area of a circle sector (wedge) can be found using the ratio
  sector area : circle area = arc degrees : 360

Other Geometry Facts
- The area of a rectangle is equal to the base times the height.
- Parallel lines cut by a transversal have equal alternate interior angles.
- A cylinder is any solid with all cross-sections parallel to the base the same shape and size.
- The volume of any shaped cylinder is equal to the area of the base times the height.
- The surface area of a cylinder is equal to the height times the perimeter of the base.

Problem Solving Strategies
- Always draw and label a picture corresponding to the information in the problem.
- Determine which quantities are known and which are being sought; this

will help you decide which formula to use, especially with trigonometry.

- Be particularly careful with angles of depression. Mislabeling them is one of the primary sources of error in these problems.

- Always check to be sure that you have answered all the questions asked by the problem.

Summary Exercises

1) From a 400 foot lighthouse a ship is seen. The angle of depression to the ship from the lighthouse is 3.64°. How far is the ship from the lighthouse?

2) An airplane takes off from sea level with an airspeed of 265 ft/sec and climbs at an angle of 8.7° with the horizontal until it reaches an altitude of 5800 feet. How long does it take the plane to reach this altitude?

3) When visibility is poor, some airplanes use an automatic landing system governed by computers. Lock-on for a certain system occurs when the airplane is 4.0 miles (slant distance) from the runway and at an altitude of 3800 feet. If the glide path is a straight line to the runway, what angle does the path make with the horizontal?

4) A train travels 2.50 miles on a straight track with a grade of 1.1°. What is the vertical rise of the train in that distance, to the nearest foot? (1 mi = 5280 ft.)

5) A girl flying a kite holds the string 3 feet above the ground. The string of the kite is taut and makes an angle of 80° with the horizontal. Approximate the height of the kite above level ground if 210 feet of string is let out.

6) The length of each blade of a pair of scissors from the pivot to the point is 6.0 inches. When the scissors are open to where the points are 4.0 inches apart, what angle do the blades make with each other?

7) What is the smallest diameter log that can be cut into a square post measuring 10 inches on a side?

8) A 24-inch-wide sheet of aluminum is bent along its center line to form a V-shaped gutter. Find the angle needed to make the gutter 7 inches deep.

9) A conveyor belt to carry materials between two levels will be installed in a factory. If the optimum angle of elevation of the belt is 22° and the distance from the first floor to the second floor is 18 feet, how long should the conveyor belt be?

10) Find a solution to the field goal problem outlined on page 56. How much better is the angle at the 10 yard line? Do you think that the 30 yard line angle of success can become better than the 10 if the kicker isn't in the middle of the field? Suppose the kicker is on the hash mark (the farthest off center that he can be). Can you find the angles of success for this position if the hash marks are 53 and 1/3 feet apart?

11) We are building a circular in-ground swimming pool and need to pour a concrete foundation for the sides and bottom. If the pool is 30 feet across and a uniform 5 feet deep, the concrete walls need to be 3 inches thick, and the

concrete floor needs to be 5 inches thick, how many cubic yards of concrete will we need?

12) We are building a circular in-ground swimming pool and need buy a vinyl liner for the sides and bottom. If the pool is 30 feet across and a uniform 5 feet deep, how many square feet will we have to cover with vinyl? If we can have clean water delivered at $50 plus an additional 15 cents for each gallon, how much will the water cost? (One gallon of water is about 0.16 cubic feet.)

13) The Anderson's want to renovate their beach cottage. They plan on repainting the exterior siding and re-shingling the roof. The northern and eastern exposures are shown in the drawing to the right. The northern and southern exposures have two 4 by 4 foot windows. The western side has no windows (just a door that also needs painting), and the eastern side has a triangular sky light and a large sliding glass door as shown in the picture.

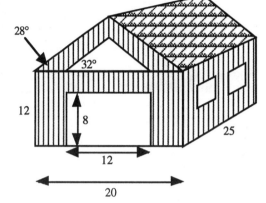

a) How much paint should be purchased if they want to put one coat on all of the siding, and each can of paint covers 600 square feet?

b) How many shingle packets should be bought to re-shingle the entire roof if each packet covers 25 square feet?

14) One way of determining the area of an irregularly shaped region is to divide it into triangles and add up all of the triangle areas. Estimate the area of the irregularly shaped region given in the picture to the right from the various measurements provided.

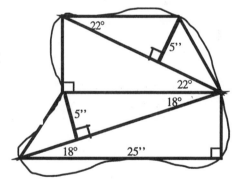

15) Consider the plan for the family area of a house as pictured, where length measurements are in feet. Calculate the cost of carpeting the floors in all the rooms, if
• carpet is $17.00 per square yard.
• padding is $2.50 per square yard.
• installation is $1.00 per square yard.

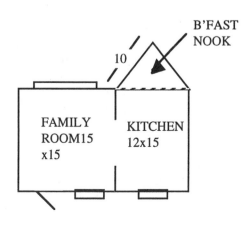

16) The Seaside Condominiums at Myrtle Beach have placed a sea wall 5 feet high between the buildings and the ocean to protect them from water damage during storms. The sea wall is 75 feet from the usual high tide. The beach slopes from the sea wall down to the shore at an angle of 3°.

a) How much higher than the usual tide will the ocean have to be before there is no beach at all, i.e., before the water reaches the sea wall?

b) There are reports that global warming due to the greenhouse effect is causing the polar ice caps to melt, raising the level of the ocean. If the oceans are rising at a rate of 1.9 inches per year, how long will it be before there is no beach at the Seaside? How long before the high tide will regularly be over the sea wall?

2   FINANCE

Sooner or later everyone will encounter **interest**. From opening a savings account to financing tuition or a car, most monetary transactions involve interest. Interest may be something you pay when you borrow or something you get paid when you invest. The calculation of interest is actually straightforward, and a few formulas will handle a multitude of situations. We will learn about these formulas and how to apply them in this chapter.

After your work on finance, you will know how to apply the formulas presented in the chapter and how to evaluate the formulas using your calculator. You will use a spreadsheet program to create amortization tables for loans and to analyze other financial situations. You also will learn how to apply this knowledge to your own personal finances.

The History of Interest and Banking

The word **interest** comes from the Latin *Id quod inter est* or "that which is between." Charging interest for the use of money has been around since Babylonian times (2000 BC). A 1700 BC cuneiform tablet has been found which contains the problem of determining how long it will take an amount of money to double if interest is compounded annually at a certain rate. Interest rates have varied widely throughout history. Babylonian rates were 20%, and rates in Cicero's Rome reached 48%. During the time of Justinian (483 - 565) rates were limited by law to 6%. By 1304, interest rates had risen to an incredible 220%. Today's interest rates are based upon current financial conditions in the world and thus are subject to change on a daily basis.

There are essentially two situations in which interest is applied.

- **saving/investing money:** The investor is paid interest on the amount invested, called the principal, usually at specified time intervals. Stocks, bonds, interest earning bank accounts and savings plans all fall into this category.

- **borrowing money:** The borrower pays the lender interest on the amount owed, also called the principal, usually at specified time intervals. Note that purchasing an item on credit also fits into this category.

These two situations actually give rise to three different formulas, investments made in one lump sum deposit, investments made with regularly scheduled deposits, and loans with regularly scheduled payments.

PERCENT, PROPORTION AND GROWTH   1

Most students learn to use percents in middle school, but many people still have trouble interpreting information involving percents even as adults. Consider the following situation.

In 1992 WinComp Computer, Inc., controlled 27% of the U. S. personal computer market; in 1993 its share of the market fell by 4%. If the annual sales of computers in the U.S. is given in the table below, by what percent did the <u>number</u> of computers sold by WinComp change from 1992 to 1993?

YEAR	Unit Sales to Dealers
1992	4,875,000
1993	5,850,000

This question states change in market share as a percent, and asks for change in number of computers sold. We have an added glitch in that the total number sold is different in the two years. We'll need to do several calculations involving percents to answer this question.

Before solving this problem, let's review the concept of **percent (%)**. The word percent comes from the Latin *per centum* meaning roughly "out of one hundred," so we can think of 22% as "22 out of 100." Thus a percent is a symbol representing a ratio or relative size. Percents must be changed to fraction or decimal numbers before we can compute with them. The ratio 22 : 100 gives:

$$\frac{22}{100} = 0.22.$$

We will most often use the decimal form. From this example, we see that the mechanics of changing a percent to a decimal number involves moving the decimal point two places to the left. The percent symbol can help us remember this:

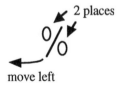

2 places

move left

Percents are often used when representing proportions and growth. We need to understand expressions like:

What is 20% of 35?
11 is what percent of 20?
32% of what number is 80?

15% markup
25% discount
6.75% interest

The first three deal with proportions (equal ratios) and the second three with growth (or decrease). We'll look at these expressions in the following examples.

Example 1.1: Answer the following questions: What is 20% of 35? 11 is what percent of 20? 32% of what is 80?

> Solution: "**What is 20% of 35**" represents the proportion
> 20 : 100 = ? : 35, or
> $$\frac{20}{100} = \frac{?}{35}.$$
>
> Notice that this is the same as 0.20 x 35 = ? after some simplification. This leads us to the notion of translating 'of' as 'times' and 'is' as 'equals.'
> Thus 20% of 35 = 0.20 x 35 = 7.
>
> Using this idea on "**11 is what percent of 20**" gives 11 = ? x 20. Does this seem correct from the ratio point of view? Check it; it works. Usually a variable such as x is used rather than a question mark, leaving us with a simple equation to solve:
> $$11 = 20x$$
> $$0.55 = x$$
>
> So 11 is 55% of 20. (Remember to move the decimal.)
>
> Now "**32% of what number is 80**" is easy: $0.32x = 80$
> $$x = 250$$

The second three expressions deal with increasing or decreasing a quantity by some percentage. In retail sales a markup is an increase in price and a discount is a decrease in price. We've already said that interest might be something we earn (an increase in our money) or something we pay (a decrease).

Example 1.2: A department store buys their camping equipment from a sporting goods wholesale firm and sells it to the general public after a **15% markup**. Given the wholesale price list below, calculate the retail price for each item.

camping equipment	2 person tent	propane lantern	first aid kit	air mattress	sleeping bag	back pack
wholesale price	$44.55	$25.49	$9.99	$17.50	$31.50	$29.97

> Solution: A 15% markup is 15% of the original amount added on to the original amount. So we need to calculate the increase and then add:

Tent:	$0.15 \times 44.55 + 44.55 = 6.6825 + 44.55 = 51.23$
Lantern:	$0.15 \times 25.49 + 25.49 = 3.8235 + 25.49 = 29.31$
Kit:	$0.15 \times 9.99 + 9.99 = 1.4985 + 9.99 = 11.49$
Mattress:	$0.15 \times 17.50 + 17.50 = 2.625 + 17.50 = 20.13$
Bag:	$0.15 \times 31.50 + 31.50 = 4.725 + 31.50 = 36.23$
Back Pack:	$0.15 \times 29.97 + 29.97 = 4.4955 + 29.97 = 34.47$

Look closely at the calculation for the retail price of the tent in Example 1.2, and regroup the terms. We get:

$$\begin{aligned} \text{Retail} &= 0.15 \times 44.55 + 44.55 \\ &= 44.55(1 + 0.15) \\ &= 44.55(1.15) = 51.23 \end{aligned}$$

So another way to calculate a 15% markup is as $1.15 \times$ the base price. Markdowns work similarly.

Example 1.3: The department store decides to get out of the retail camping equipment business and has a clearance sale: The advertisement says: "*15% Discount* off of retail on all Camping Equipment." Calculate the sale prices for the equipment from Example 1.2.

Solution: A 15% discount is 15% of the original amount subtracted from the original amount. So we need to calculate the discount and then subtract:

Tent: $51.23 - 0.15 \times 51.23 = 51.23(1 - 0.15) = 51.23(0.85) = 43.55$

So a 15% markdown is the same as multiplying by $(1 - 0.15)$. Let's do the rest this way.

Lantern:	$29.31(0.85) = 24.91$
Kit:	$11.49(0.85) = 9.77$
Mattress:	$20.13(0.85) = 17.11$
Bag:	$36.23(0.85) = 30.80$
Back Pack:	$34.47(0.85) = 29.30$

It is interesting to note that the 15% discount does not "undo" the 15% markup; this is a common misconception. In fact when we marked up the items 15% and then down 15%, we ended up with a lower price than we started with.

Now we're ready to solve the WinComp computer problem. Let's state it again.

Example 1.4: In 1992 WinComp Computer, Inc., controlled 27% of the U. S. personal computer market; in 1993 its share of the market fell by 4%. If the total annual sales of computers for all companies in the U.S. is given in the table below, by what percent did the <u>number</u> of computers sold by WinComp change from 1992 to 1993?

YEAR	Unit Sales to Dealers
1992	4,875,000
1993	5,850,000

<u>Solution:</u> Let's check first to see what the actual numbers of computers sold by WinComp were for 1992 and 1993.

1992: A 27% market share implies that they <u>sold</u> $0.27 \times 4,875,000 =$ 1,316,250 computers.

1993: A 23% market share implies that they <u>sold</u> $0.23 \times 5,850,000 =$ 1,345,500 computers.

They lost market share but they sold 29,250 *more* computers! What was the percent change in the number of computers sold from 1992 to 1993?

$$\frac{\text{change}}{\text{original}} = \frac{29,250}{1,316,250} = 0.022 \text{ or a } 2.2\% \text{ increase.}$$

YOU TRY IT 1.1

The Morning Glory Daycare Center is having an auction to raise money for new swing sets. They have received several items at reduced cost to auction off. If the center wants to earn at least a 5% profit, what should it set the minimum bids to for the following items?

item	microwave	television	car stereo	Alaska cruise
cost	89.00	249.00	68.88	2,105.00

YOU TRY IT 1.2

A manufacturer of metal skis produces three models: Standard, Professional and Competition. Market analysis has revealed that the retail prices given below will be acceptable to the skiing public. The accounting department needs to know the actual cost to the factory for each model. Calculate this cost.

Model	Retail Price	Mark-up over factory cost
Standard	$103.50	15%
Professional	$126.50	15%
Competition	$174.00	20%

The last thing we need to look at before beginning our study of finance is how to handle interest as a percent. Suppose that we buy $10,000 worth of stock that has an expected return of 6.75% interest in a year. The company's payment policy is to send stock holders an annual dividend check. Assuming that the actual earnings are 6.75% per year, how much will the dividend check be?

To figure the interest, we need to calculate what 6.75% of 10,000 is. We can use our standard translations for 'of' and 'is' to get:

$$0.0675 \times 10000 = \$675.$$

In general, **interest earned in one period** can be calculated as:

$$\boxed{\begin{array}{c} \textbf{Simple Interest} \\ \textbf{I = r P} \end{array}}$$

where **r** is the interest rate and **P** is the amount earning interest, often called the principal.

Example 1.5: How much will we have earned from our $10,000 in stocks over 10 years if the dividend rate stays the same?

Solution: We earned $0.0675 \times 10000 = \$675$ from one year, and unless we buy more stock or the dividend rate changes, we'll earn $675 \times 10 = 6750$ or $6,750 in 10 years.

YOU TRY IT 1.3

Suppose that we buy $100,000 stock in a company that promises a minimum return of 2 percent in the first year. How much interest will we have earned in the first year?

YOU TRY IT 1.4

It is common today for an individual to have a stock portfolio, rather than putting all her shares in one company. If a portfolio consists $1000 each of the following stocks (stated together with their minimum expected annual returns), what will the annual income from the portfolio be?

Company	Return	
Pacific Power	11.6%	116.00
Sparkle Beverages	9.2%	92.00
Iron works of Poland	5.25%	52.50
Tendyne Computers	7%	70.00
PS Health Care Associates	8%	40.00

410.50

This situation -- interest paid to the investor at regular intervals on the amount invested -- is called **simple interest**. Bank accounts don't usually work this way. The interest is deposited into the account so the amount of money earning interest grows bigger. This more common way of accruing interest is called **compound interest**, and is the subject of the next section.

Exercise Set 2.1

1) a) What is 25% of 188? 47% b) What percent of 80 is 66? 82.5%
   c) 16% of what number is 35? 218.75

2) a) What is 80% of 208? $80 \cdot 208$ b) What percent of 44 is 27? $27/44 = 61.36$
   c) 8% of what number is 122?

3) a) What is 75% of 110? $75 \cdot 110$ b) What percent of 56 is 30? $56\overline{30}$ 53.57% $30/56$
   c) 12% of what number is 35?

4) a) What is 3% of 1,024?   b) What percent of 1,000 is 245? 24.5%
   c) 11% of what number is 15?

5) A store is having a 10% off sale on all merchandise. What will the following items cost on sale: A designer sweater that retails for $149.00, a leather wallet that retails for $49.99, a man's suit that retails for 799.00?

6) A building supply store manager gets the bright idea to mark up selected items 50% and then have a 50% off sale, thinking he can come out even. How much will he lose on the sale of a $4600.00 hot tub?

7) Greg and Marsha hired a contractor to build their house for "cost plus 10," i.e., the cost of materials and labor plus 10%. Another contractor offered to build the house for a flat rate of $122,000.00, but Greg chose not to use him. If the cost of materials and labor totaled $111,432.26, how much did Greg and Marsha pay? Would they have been better off with the flat rate builder? Why?

8) A builder borrowed $21,000 for one year to purchase equipment. If he paid $2225 in simple interest, what percent interest did he pay?

9) Of the 87 senior mathematical science majors at a particular institution, 45 are specializing in secondary education, 37 are specializing in computer science 4 are specializing in statistics, and the rest are either pure math or applied math majors. What percent of the mathematical science majors are specializing in:

   a) computer science?   b) statistics?

   c) secondary education?   d) math or applied math?

10) An attorney wishes to purchase a building in which to locate her office. She finds two buildings, one within city limits and one out in the county. She was warned that the real estate taxes are calculated differently in the two locations. The city charges 400 mills ($4.00 per $1, 000 of assessed value)

with assessment at 90% of market value. The county charges 460 mills ($4.60 per $1,000 of assessed value) with assessment at 50% of market value. If the properties have market values of $125,000 each, what are the taxes on the two buildings?

11) A developer has a 92.5 foot wide lot on which he wants to build a house. The local building codes require that 15% of the width of the lot be reserved for side yards and that the house be no closer than 10 feet to any property line. Is the lot wide enough for a 59 foot wide house?

12) Each time a song is played on the radio, the record company and the songwriter are paid a royalty of $0.30. Of the total, 75% goes to the company and the rest to the writer. If on a network of 50 radio stations, a certain song is played 4 times a day during the first week and then 20 times a day for the next three weeks, how much does the network owe in royalties for the four weeks? How much do the record company and the songwriter receive each?

A fundamental knowledge of basic proportions and percents is very helpful on the Graduate Record Examination (GRE). The questions below are similar in structure and content to those appearing in the quantitative section of the GRE. The directions for problems 13 through 15 are: Each of the following problems contains two quantities labeled A and B. Determine the relationship between the sizes of the quantities. Answer (A) if A > B, (B) if A < B, (C) if A = B, or (D) if the relationship between A and B cannot be determined. **Explain your reasoning.**

13) A: 6% of 9   B: 9% of 6

14) A hardware store purchased hammers for $7 each and sold them for 25% above cost.

A: $9.10   B: The price at which the store sold a hammer.

15) A: 30% of 50   B: $\frac{3}{10} \times 50$

The GRE also contains multiple choice questions similar to problems 16 through 20 below. Explain the reason for your answer.

16) 0.75% =

a) $\frac{3}{4}$ b) $\frac{3}{40}$ c) $\frac{3}{400}$ d) $\frac{3}{4000}$ e) $\frac{3}{40000}$

17) $\frac{3}{4}\%$ =

a) $\frac{3}{4}$ b) $\frac{3}{40}$ c) $\frac{3}{400}$ d) $\frac{3}{4000}$ e) $\frac{3}{40000}$

18) A widow received 50% of her husband's estate, and each of her four sons received 25% of the balance. If the widow and one of her sons received a total of $100,000 from the estate, what was the amount of the estate?

a) 125,000   b) 140,000   c) 160,000   d) 200,000   e) 300,000

19) The estate of a retired banker amounted to $750,000. The will gave the widow $300,000 and divided the rest evenly between 3 children. What percent of the estate did each child receive?

    a) 15%   b) 18%   c) 20%   d) 22%   e) 25%

20) If $1000 is deposited into an account paying 5 percent interest in one year, how much interest is earned?

    a) $5   b) $50   c) $100   d) $250   e) $500

- SAVING AND INVESTING MONEY - COMPOUND INTEREST

2

Compound interest means that interest is paid on the **principal** (the amount deposited or invested) as well as the interest that has already accumulated. For example, if $1000 is deposited at 6% interest compounded annually, then the interest for the first year is

$$(0.06) \times (1000) = \$60.$$

Therefore the account balance at the end of the first year is $1000 + $60 = $1060. During the second year, interest is paid on the entire $1060 so the interest earned at the end of the second year is

$$(1060) \times (0.06) = \$63.60,$$

and the amount in the account at the end of the second year is

$$\$1060 + \$63.60 = \$1123.60.$$

How would we find the amount in the account after 10 years? We need a formula. Let's consider what was done when the amount in the above account was calculated. Let **A** stand for the amount in the account after interest is earned (sometimes called the **future value**), **P** stand for the principal, and **r** stand for the annual interest rate. Then

interest earned in one year = P x r.

We found the new amount in the account by adding this interest to the principal. After one year, $A = P + P \times r$ or, factoring, $A = P(1 + r)$. This final amount would then be used as the principal for the second year. The interest earned the second year will be $P(1 + r)r$, since the account had $P(1 + r)$ in it at the start of year 2. So adding the interest earned to the principal for the second year, we get

$$A = P(1 + r) + P(1 + r)\, r = P(1 + r)(1 + r) = P(1 + r)^2.$$

This gives us the following table:

Year	Amount in the account (principal + interest)		
0	$P + 0$		
1	$P + P \, r = P(1 + r)$		
2	$P(1 + r) + P(1 + r) \, r$	$= P(1 + r) \, (1 + r)$	$= P(1 + r)^2$
3	$P(1 + r)^2 + [P(1 + r)^2] \, r$	$= P(1 + r)^2 \, (1 + r)$	$= P(1 + r)^3$
4	$P(1 + r)^3 + [P(1 + r)^3] \, r$	$= P(1 + r)^3 \, (1 + r)$	$= P(1 + r)^4$

Notice that for each year, the exponent is the same as the number of years. This pattern leads to a formula to find the amount in an annually compounded interest account at the end of a given time period. Let **n** be the number of time periods:

> **Lump Sum Formula**
> $$A = P(1 + r)^n$$

Example 2.1: Melissa deposits \$15,000 in an account paying 8% interest compounded annually. How much will she have in the account after 6 years?

Solution: Using the formula: $A = 15{,}000(1 + 0.08)^6$.

To calculate these kinds of expressions on a calculator, always start with what is in the parentheses. In this case enter $1 + 0.08$ which equals 1.08. Next raise 1.08 to the sixth power. (Your calculator has a power key that looks like y^x, x^y or ^.)

Enter 1.08 y^x 6 = to get 1.586874323. (Note: Don't round yet.)

Lastly, $A = 15{,}000(1.586874323) = 23{,}803.11484$. Since we are working with money, we round our answer to two decimal places. Therefore Melissa will have \$23,803.11 in this account at the end of six years.

To calculate the amount of **interest** that Melissa's money earned in the six years, we would merely subtract the original deposit (principal) from the final amount in the account:

Interest	= Amount - Principal
I	$= A - P$
I	$= 23{,}803.11 - 15{,}000$
I	$= \$8{,}803.11$

A note on rounding: Since we are dealing with money, all final answers should be rounded to cents. BUT: Don't round too soon, especially if the problem involves a long time period. Small errors (like rounding) tend to add up over a long time period. We will be keeping ALL digits given by the calculator in the intermediate steps of these problems. Most calculators have a memory storage and retrieval feature; this can be useful. Later we'll look at a few examples where storing a number on the calculator will be advantageous.

YOU TRY IT 2.1

George deposits $10,000 in an account paying 9% compounded annually.
 a) How much will he have in the account after 5 years?
 b) How much more would he have if he could get 9.5% interest?
 c) How much interest will he earn in each case?

Multiple Compoundings

In the above examples interest was assumed to be compounded annually (once a year). However, anyone with a savings account knows that interest is deposited into the account (or **compounded**) more than once a year. Some common compounding periods are:

semiannually	twice a year
quarterly	four times a year
monthly	twelve times a year
daily	usually 365 times a year

Government regulations require that interest rates be stated as annual interest rates along with the compounding period. To work with these compounding periods, we have to adjust both the interest rate and the number of time periods. For example, suppose we have an interest rate of 6% that is compounded **monthly**. The government required <u>annual</u> rate is 6%. The actual <u>monthly</u> rate is $0.06/12 = 0.005$. If the total time were five years, then the number of time periods (usually called compounding periods) would be $5 \times 12 = 60$.

Example 2.1 Again: Suppose that the interest rate of 8% were compounded **quarterly**. Then the quarterly interest rate would be $0.08/4 = 0.02$, which we use for **r** in the formula. Also, over the course of six years, interest will be compounded $6 \times 4 = 24$ times so this becomes **n** in the formula. The principal is still $15,000; no adjustments are necessary for the principal.

We use the same formula again: $$A = P(1+r)^n$$

A = future value of account (amount in the account after n compounding periods)

P = principal (amount deposited)

r = periodic interest rate (rate for one compounding period, calculated as annual interest rate divided by number of compounding periods)

n = term (total number of payments made, calculated as number of years times the number of compoundings)

Solution: Using the formula: $A = 15,000(1 + 0.02)^{24}$

Enter 1.02 raised to the 24th power to get 1.60843725.

We need to multiply this (already on your calculator screen) by 15,000: Press **x 15000 =** to get 24126.558.

Melissa will have $24,126.56 in this account at the end of six years with quarterly compounding.

NOTE: The approach used above to calculate the total in the account on a calculator is not the only way to arrive at the correct solution. If your calculator has parenthesis buttons -- (and) -- you can type it as it is written. In the next example, we take that approach.

Example 1.2: Jerry deposits $1200 in an account paying 9% interest compounded monthly. How much money will he have in the account after 20 years?

$1200(1+.0075)^{240}$

Solution: The principal is the amount deposited: **P** = 1200

We must adjust rate and term for monthly compoundings:
   r = 0.09/12 = 0.0075, and **n** = 12 x 20 = 240
   (Since we are compounding twelve times per year.)

Therefore   **A** = $1200(1 + 0.0075)^{240}$

Typing this into the calculator as 1200 x (1 + 0.0075) (power key) 240 gives
          = $7,210.98 (rounded.)

YOU TRY IT 2.2

Alice deposits $850 into an account paying 12% compounded quarterly.
a) How much will be in the account after 10 years?
b) How much more would be in the account if the account paid 12.5%?
c) How much interest does each one earn?

$850(1+.03)^{40}$

Periodic Payments

In the above compound interest problems, we were taking one lump sum and depositing it up front all at once. More often, payments are made into the account at the end of **regular time intervals**. If the frequency of payments matches the frequency of compounding (as it will in our problems), accounts with periodic payments are called **ordinary annuities**. Suppose we deposit $50 at the end of each year for five years into an account paying 6% interest compounded annually. Let's calculate how much money would be in the account at the end of the five years.

The first deposit of $50 will be in the account for four years, since it will be deposited at the end of the first year, so using our formula for compound interest

$$A = 50(1 + 0.06)^4 = \$63.12.$$

The second deposit of $50 will be in the account for three years so it's future value will be

$$A = 50(1 + 0.06)^3 = \$59.55.$$

The third deposit will yield

$$A = 50(1 + 0.06)^2 = \$56.18,$$

and the fourth deposit will yield

$$A = 50(1 + 0.06)^1 = \$53.00.$$

Since the last deposit will be made at the end of the fifth year, it will not earn any interest so its value is just $50. Adding these gives

$$50 + 50(1 + 0.06)^1 + 50(1 + 0.06)^2 + 50(1 + 0.06)^3 + 50(1 + 0.06)^4 = \$281.85$$

This method would become tedious if these deposits were made for many years, so we really need a formula. Let **A** stand for the amount in the account and **r** represent the interest rate per compounding period. Also let **R** stand for the amount of each payment being made. Using the above example, we can conclude that the future amount, **A**, of payments, **R**, after five years can be represented by:

$$A = R + R(1 + r) + R(1 + r)^2 + R(1 + r)^3 + R(1 + r)^4$$

We now perform some algebra "trickery" to get a simpler formula. First multiply this equation on both sides by $(1 + r)$:

$$A(1 + r) = R(1 + r) + R(1 + r)^2 + R(1 + r)^3 + R(1 + r)^4 + R(1 + r)^5$$

Subtracting the original equation from the new equation yields:

$$\begin{aligned} A(1 + r) \quad &= \quad R(1 + r) + R(1 + r)^2 + R(1 + r)^3 + R(1 + r)^4 + R(1 + r)^5 \\ - [\quad A \quad &= R + R(1 + r) + R(1 + r)^2 + R(1 + r)^3 + R(1 + r)^4 \qquad] \end{aligned}$$

Notice that many of the terms are the same and will subtract out yielding:

$$A(1 + r) - A = - R + R(1 + r)^5$$

A little more algebra gives: $A = \dfrac{R[(1 + r)^5 - 1]}{r}$

Notice that the exponent is the same as the number of compounding periods, so if we let **n** stand for the number of compounding periods as before, then the general formula would be:

$$\boxed{\begin{array}{c} \textbf{Periodic Payment Formula} \\[2mm] A = \dfrac{R\left[(1+r)^n - 1\right]}{r} \end{array}}$$

where **R** is the periodic payment, **r** is the rate per compounding period, and **A** is the amount in the account after **n** compoundings (A is often called the future value of an ordinary annuity). Now that we have a formula to work with, let's look at some examples.

Example 2.5: If $100 is deposited in an account at the end of each year for 20 years at 7.25% interest compounded annually, how much will be in the account at the end of twenty years?

Solution: **R** = 100, **r** = 0.0725 and **n** = 20.

Thus
$$A = \frac{100[(1 + 0.0725)^{20} - 1]}{0.0725}$$

First add what is in the parentheses, then raise it to the 20th power.
$$A = \frac{100[4.054581338 - 1]}{0.0725}$$

Now do the subtraction in the brackets.
$$A = \frac{100(3.054581338)}{0.0725}$$

Multiply what is in the numerator.
$$A = \frac{305.4581338}{0.0725} = \$4213.22$$

So, there will be $4,213.18 in the account at the end of twenty years.

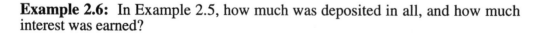

Example 2.6: In Example 2.5, how much was deposited in all, and how much interest was earned?

Solution: Since $100 is deposited at the end of each year for 20 years, the total deposit is (100)(20) = $2000.

Since the total value at the end of the 20 years is $4213.18, the difference between the total value and the total amount actually deposited is the amount of interest the money earned:

$$4213.18 - 2000 = \$2213.18.$$

YOU TRY IT 2.3
Ellen gets a bonus at the end of each year at her company. If she deposits $850 at the end of each year into an account paying 6.8% compounded annually,
a) How much will she have after 15 years?
b) How much of this amount was deposited and how much is interest?
c) Using the formula find how much she will have in the account after one year. Explain why this makes sense.

Example 2.7: If $75 is deposited at the end of each month in an account paying 6% monthly, what will the future value of the account be at the end of three years?

Solution: **R** = 75. We have to make the usual adjustment in the interest rate and the number of compoundings, so **r** = 0.06/12 = 0.005, and **n** = 3 x 12 = 36. So,

$$A = \frac{75[(1 + 0.005)^{36} - 1]}{0.005} = \frac{75[0.1966805]}{0.005} = \$2,950.21$$

Example 2.8: Paul needs $4,000 in two years to go to Australia. He plans to make equal payments at the end of each month into an account paying 9% interest compounded monthly. How much should his payments be?

Solution: We are looking for the amount of the payments, **R** in the formula. So **A** = $4,000, **n** = 2 x 12 = 24, and **r** = 0.09/12 = 0.0075.

Plugging these into our formula gives:

$$4,000 = \frac{R[(1 + 0.0075)^{24} - 1]}{0.0075}$$

Multiply both sides by 0.0075

$$30 = R[(1 + 0.0075)^{24} - 1]$$

Calculate what is in the brackets

$$30 = R(0.196413529)$$

Now divide

$$152.74 = R \quad \text{(rounded)}$$

Paul must deposit $152.74 at the end of each month to save the $4,000.

YOU TRY IT 2.4

Andy is saving for a summer trip to Europe. If he deposits $110 at the end of each month into an account paying 8.75% compounded monthly,

a) How much will he have in his account after two years?

b) How much of this total value is actually deposited and how much is interest?

c) How much would he have in the account after $2\frac{1}{2}$ years?

YOU TRY IT 2.5

After finishing college, Joe starts his new job. Immediately he starts depositing some of each end-of-month paycheck into an account paying 6% compounded monthly. How much should he deposit at the end of each month in order to have a $2500 down payment on his dream car,

a) in two years? How much does he deposit, total?

b) in four years? How much does he deposit, total?

c) Since two years is exactly half of four years, why isn't the total in part b exactly half of the total in part a?

Present Value

Sometimes the question we want to answer isn't "Given so much now what will we have later?", but rather "How much do we need now to have some needed amount later?" So we're given the future value and asked about the **present value**. For lump sums this is easy:

Example 2.9: How much money must be deposited into an account paying an interest rate of 12% compounded semiannually, so that $15,000 can be withdrawn in 10 years? (This can also be asked as "what is the present value of $15,000 due in ten years?")

Solution: We are asked to find the amount deposited, which is **P** in the formula. Therefore **A** = 15,000, **r** = 0.12/2 = 0.06, and **n** = 2 x 10 = 20. (We divided by two to get **r** and multiplied by two to get **n** since we are compounding twice a year!)

$$15,000 = P(1 + 0.06)^{20}$$

$$15,000 = P(3.207135472)$$

$$P = \frac{15000}{3.207135472} = \$4,677.07$$

YOU TRY IT 2.6

John plans to buy a $4000 used car three years from now. His bank offers 4% compounded daily on savings accounts.
a) How much must he deposit today to have the $4000 in 3 years?
b) Suppose another bank pays 5% but charges $10 to open a new account. Would John be better off opening a new account or staying with his own bank? Explain.

The present value of an ordinary annuity (periodic payments) is more complicated, but we have all the necessary tools: Think of the present value of an ordinary annuity as: "What size lump sum deposit must be deposited today in order to have the same amount as the future value of a given annuity?"

Example 2.10: The Gordons start a retirement annuity in which they deposit $100 per month at 6% interest compounded monthly. They plan on continuing their $100 monthly deposits for 30 years. What is the present value of this annuity?

Solution: Finding the present value for periodic payments is the same as finding the lump sum that would yield the same final amount **A.** This is a two step process.

1) Find the amount in the annuity after 30 years: Plugging the given information into our formula gives:
$$A = \frac{100[(1 + 0.005)^{360} - 1]}{0.005}$$
$$A = \$100,451.50$$

2) What lump sum gives the same amount?

$$\$100,451.50 = P(1 + 0.005)^{360}$$
$$P = \$16,679.16$$

YOU TRY IT 2.7
Find the present value of each ordinary annuity given below:
a) Payments of $25 are made quarterly for 10 years at 5% coumpounded quarterly.
b) Payments of $500 are made annually for 25 years at 5% compounded annually.

Calculating the Time Needed

Often we are interested in finding out how long it would take to save a particular amount of money. This translates mathematically into finding **n** in the formula. Unfortunately, **n** is in the exponent and so simple algebra is not helpful. There are two techniques for finding **n**: Trial-and-error, or a simple **logarithm** trick. Let's consider an example.

Example 2.11: Suppose $5,700 was deposited into an account paying an interest rate of 9% compounded monthly. How long would it take for the account to accumulate $7,000?

Solution: We need to use the monthly interest rate which will be $0.09/12 = 0.0075$. Also we know that **A**=7,000 and **P**=5,700. Plugging into our formula we get

$$7,000 = 5,700(1 + 0.0075)^n;$$

Dividing both sides by 5,700 we get the simplified equation:

$$1.228070175 = (1.0075)^n$$

The unknown, **n**, is in the exponent. Let's get a "ball park" figure for **n** using trial and error. Consider the chart below:

n	1.0075^n	Greater or Less Than 1.228070175
10	1.077582545	Less (a lot)
26	1.214427031	Less (closer!)
27	1.223535233	Less (very close!)
28	1.232711748	Greater

This tells us that the number of months it would take to get our $7,000.00 is between 27 and 28. We never want too little, so we'll give 28 as our answer. This approach is perfectly good, but the process can be time consuming especially if we have no notion about the size of **n**.

Can we find **n** without guessing? To solve when we have a variable in the exponent we must use **logarithms**. The definition of a log (base 10) is:

log(x) is the power of 10 that is equal to x

or

$y = \log(x)$ if and only if $x = 10^y$.

There are several useful properties of logs; we need them to answer questions like:

- What power of 10 is equal to 100? Find n: $10^n = 100$.
- What power of 2 is 16? Find n: $2^n = 16$.
- What power of 1.0075 is 1.228070175? Find n: $1.0075^n = 1.228070175$.

The first two questions should be easy to answer. 10^2 is 100 and 2^4 is 16. But the last one is harder. (It's the question we answered in Example 2.11, so we already know the answer is between 27 and 28.) We need two properties of logs to answer these questions:

We can "log" both sides of an equation: **x = y if and only if log(x) = log(y)**

We can pull a power from inside a log to down in front: $\log(a^b) = b \times \log(a)$

By taking the log of both sides of the equation and then pulling the power down in front, we can restate the first two questions again as:

- Find n: n x log(10) = log(100). Thus n = log(100) / log(10)
- Find n: n x log(2) = log(16). Thus n = log(16) / log(2)

Look on your calculator and find a key that says **LOG**. We can take the log of a number using the calculator by typing the number and then pressing **LOG** (or on a graphing calculator press **LOG** then the number). Try this for the two calculations above:

- n = log(100) / log(10) = 2 / 1 = 2 (as expected)
- n = log(16) / log(2) = 1.204119983 / 0.3010299957 = 4 (as expected)

Example 2.12: Find the following exponents using logs.

$$\text{a)} \quad 1.25 = 1.0034^n \quad \text{b)} \quad 2.1 = \left(1 + \frac{.05}{12}\right)^n \quad \text{c)} \quad 100 = \frac{20\left((1.03)^n - 1\right)}{.03}$$

Solution:

a) Taking the log of both sides and moving the exponent down gives:

$$\log(1.25) = n \times \log(1.0034) \quad \text{or} \quad n = \frac{\log(1.25)}{\log(1.0034)} = 65.74 \text{ (rounded)}$$

b) Taking the log of both sides and moving the exponent down gives:

$$\log(2.1) = n \times \left(1 + \frac{.05}{12}\right) \quad \text{or} \quad n = \frac{\log(2.1)}{\log\left(1 + \frac{.05}{12}\right)} = 178.44 \text{ (rounded)}$$

c) We can't take the log of both sides yet since the exponent factor is not isolated. We need to move the 20 and the 0.03 out of the way first. So we divide by 20 and multiply by 0.03 to move them:

$$\frac{100 \times 0.03}{20} = (1.03)^n - 1 \quad \text{or} \quad 0.15 = (1.03)^n - 1$$

We are still left with the -1. Add 1 to both sides to get:

$$1.15 = (1.03)^n$$

Now we are ready to use the log trick:

$$\log(1.15) = n \times \log(1.03) \quad \text{or} \quad n = \frac{\log(1.15)}{\log(1.03)} = 4.73 \text{ (rounded)}$$

Can you do this problem with trial and error? Try it.

YOU TRY IT 2.8
Find the exponents in the following equations.

a) $2.15 = 1.000124^n$

b) $4.1 = \left(1 + \dfrac{.075}{4}\right)^n$

c) $1000 = \dfrac{125\left((1.11)^n - 1\right)}{.11}$

Let's redo Example 2.11 using logs.

Example 2.11 (Again): Suppose $5,700 was deposited into an account paying an interest rate of 9% compounded monthly. How long would it take for the account to accumulate $7,000?

> <u>Solution:</u> From our last attempt, we have the simplified equation:
> $$1.228070175 = (1.0075)^n.$$
> Using the log properties: $\qquad \log(1.228070175) = n\,\log(1.0075)$
>
> We can divide before taking logs: $\; n = \dfrac{\log(1.228070175)}{\log(1.0075)} = 27.5$
>
> We'll round up since 27 wouldn't be quite enough time. Thus, it will take 28 monthly periods, or 2 years and 4 months to accumulate the $7,000, the same answer we got through trial-and-error!

Your calculator probably has one or two other logarithms, LN and possibly LOG$_2$. These logs behave the same as LOG except that the base number is different. LN uses the scientific number e as its base, and LOG$_2$ uses 2. You can use these buttons instead of LOG, since the two properties we used hold for all logs, but be careful not to switch between different logs in the middle of a problem; your answer will be incorrect.

Example 2.13: Suppose we deposit $200 at the end of every month into an account paying 9% interest compounded monthly. How long will we need to continue making these deposits before the account has $20,000 in it?

Solution: We are looking for the **n** in our **annuity** formula. **R** = $200, **A** = $20,000, **r** = 0.09/12 = 0.0075.

Plug into the formula:
$$20,000 = \frac{200\,[(1 + 0.0075)^n - 1]}{0.0075}$$

We need to isolate the power before using logs. To do this peel away the numbers attached to it one at a time, undoing the operations one at a time. Here is one possible sequence of operations:

First divide both sides of the equation by 200:
$$100 = \frac{[(1 + 0.0075)^n - 1]}{0.0075}$$

Multiply both sides by 0.0075: $\qquad 0.75 = (1 + 0.0075)^n - 1$

Add what is in the parentheses: $\qquad 0.75 = (1.0075)^n - 1$

Adding one to both sides yields: $\qquad 1.75 = (1.0075)^n$

At this point we could try trial-and-error:

n	1.0075^n	Greater or Less Than 1.75
28	1.2327117	A lot less
100	2.111084	Way too much
50	1.452957	Less
70	1.687151	Less (but close)
75	1.751375	More (but close)
74	1.738337	Found it!

Looks like it's between 74 and 75 months, so we'll use 75 months, since 74 would fall short. Thus, it would take 75/12 = 6 years and 3 months to have $20,000.

Alternately, we could use the log trick: $\qquad \log 1.75 = n\,(\log 1.0075)$

$$\frac{\log 1.75}{\log 1.0075} = n$$

$$n = 74.89,$$

And we arrive at 75 months again.

YOU TRY IT 2.9

Suppose $8500 is deposited into an account paying 8% compounded quarterly. How long will it take for the account to accumulate $12,000?

YOU TRY IT 2.10

Betty and John are saving for a down payment on a house. They deposit $650 at the end of each quarter into an account that pays 8.5 % compounded quarterly.

a) How long before they have the $12,000 down payment?

b) Would they have the money sooner or later than this if they put $600 per quarter in an account paying 9%? Explain your conclusion.

We have learned two formulas in this section, so let's talk about when to use which formula.

When working with a one time deposit, use:

$$A = P(1 + r)^n$$

For payments deposited at regular intervals use:

$$A = \frac{R\left[(1 + r)^n - 1\right]}{r}$$

Phrases like "monthly deposits," "every six months," or any other words that indicate a systematic and regular scheme for investing leads to the second formula. The first formula is only used when money is deposited one time. Once you decide which formula to use, notice whether you are trying to find the future amount, interest, payments, present value, etc., so you will know which variable you are solving for. Make sure your answers are reasonable. If you have a four year loan on a $10,000 car, monthly payments of $50 or $2,000 are unrealistic.

Exercise Set 2.2

1) If $2300 is deposited in an account paying an annual interest rate of 7% compounded annually for eight years, how much money will be in the account at the end of the eight years? How much interest will be earned?

2) If $6,543.21 is deposited into an account earning 7.6% compounded monthly and left there for 15 years, how much will the account be worth? How much interest will be earned?

3) If $90 is deposited into an account at the end of each year for ten years at 12% compounded annually, how much will be in the account at the end of ten years? How much is actually deposited and how much interest is earned?

4) If $5234.17 is deposited into an account at the end of each year for 13 years at 9.35% compounded annually, how much will be in the account at the end of the 13 years? How much interest is earned?

5) If $3,400 is deposited in an account paying an annual interest rate of 7.25% compounded quarterly for six years, how much will be in the account at the end of the six years? How much interest will be earned?

6) How much money must be deposited now into an account paying a rate of 9% compounded annually so that $3000 can be withdrawn in 10 years?

7) What is the present value of an account paying 8% compounded monthly that will contain $23,000 after 14 years?

8) Lucy has found an account that will guarantee her a return of 4% compounded monthly. She wants to give her newborn niece a gift for college on her seventeenth birthday. If Lucy plans to give her niece $20,000, how much must she deposit today to have the money?

9) John works part-time and earns $100 each week. He deposits his earnings at the end of each month in an account which pays 6.8% compounded monthly.

If he does this consistently for three years, will he have enough to buy the $15,000 car he's hoping to get? (Assume 4 weeks in a month.)

10) Terry wants to make equal payments at the end of each month in an account paying 11% interest compounded monthly to save $5,000 in two years to buy a boat. What should her monthly deposits be?

11) How long would it take for $12000 to double in value (be worth $24,000) if it were invested in an account which pays 5.73% compounded semiannually?

12) If $2600 is deposited in an account paying 9% compounded semiannually, how many years will it take for the money to double?

13) Fred and Kay plan to buy some land in the mountains. They need a down payment of $35,000 but they only have $25,000. They decide to invest the $25,000 in an account paying 11% compounded daily. How long will they have to wait until they have enough for the down payment?

14) If you have $800 to invest for two years, which is the better investment: 7.5% compounded annually or 7.3% compounded monthly?

15) Meg won $100,000 on a game show that she will receive in five years, when she turns 18. How much will the game show have to deposit today into an account paying 8% compounded monthly, in order to have Meg's money in five years?

16) What is the present value of $50 deposited each quarter into an account paying 7% compounded quarterly after 20 years?

17) If Karen deposits $3000 every six months into an account paying 7% compounded semi-annually, how long will it take to have $150,000 in the account?

18) George promised to pay his Dad back the $1000 he owes him at the end of next year, which is 18 months from now. He wants to take the same amount out of his paycheck each month for that time period in order to save up the money. George has a savings account that pays $5 \frac{1}{4}$% compounded monthly. How much should he deposit at the end of each month in order to have the money for his Dad on time?

19) What lump sum deposited today at 9.25% compounded semi-annually for 10 years would yield the same total amount as semi-annual payments of $350 at the same rate for 10 years?

20) Working in the U.S., Anna sends $200 at the end of every quarter to her family back home. They are secretly saving up the money to give to her so she can make a down payment on a car. If they deposit the funds at the end of each quarter in an account which earns 6.3% compounded quarterly, how long before they have the $2200 they want to give to her?

21) Ed is 38 years old and settled into a job making about $27,000 per year that he receives in equal payments at the end of each month. He decides it's time to begin putting 10% of each paycheck immediately into a retirement account which pays 6.7% compounded monthly. Alice, Ed's sister, is 22 years old. She decides to go ahead and start putting $75 at the end of each month into the same retirement account, even though she isn't making much money at her

job yet. If both continue doing the same until they retire at age 65, who will have more retirement money? Who will have deposited more?

22) Which is worth more after 10 years: $5000 deposited into an account paying 4.5% interest compounded daily (365.) - OR - Deposits of $135 at the end of each quarter into an account paying 8% quarterly.

23) After winning a sweepstakes, Laura is given the following options:
   Option 1: Receive $100,000 now.

   Option 2: Receive $3000 at the end of each quarter for the next 10 years.

   Laura finds that she can earn 5.4% compounded quarterly on any investments over the next ten years. If she doesn't need to spend any of the money, which option would be best for her? Why?

24) David deposits $500 at the end of each quarter into an account paying 6.5% interest compounded quarterly for 7 years. He then changes his deposit to $725 each quarter for 5 more years at the same rate. What will the amount on deposit be after the entire 12 years?

25) John deposited $12,800 in an account paying 7.48% interest compounded quarterly for 5 years. After the fifth year he found a better deal, so he emptied that account and deposited the total amount in an account paying 9% interest compounded semiannually. If he then left it in the second account for the next 4 years, what would be the total value of his account at the end of that time?

26) Write a response to the given memo, justifying your response.

```
To:     J. Student                              MEMO
From:   R. Stewart, Management
Re:     Planning

We need to plan for a company expansion targeted for 4 years from
now.  Please compare the two strategies listed below:
(1)   Invest $125,000 at 8% compounded quarterly
(2)   Deposit $2600 each month at 7.9% compounded monthly
```

27) Write a response to the given memo, justifying your response.

```
To:     J. Student                              MEMO
From:   R. Stewart, Alumni Association
Re:     Investments

We have the opportunity to invest a generous donation of $17,500.00
for 5 years.  Please compare our investment choices, as listed below:
(1)   6.25% compounded daily
(2)   6.4% compounded semi-annually
(3)   5% compounded monthly changing to 7% monthly after 3 years
```

The following exercises require access to a spreadsheet.

28) Create a spreadsheet in which the Principal, Compounding and Time remain the same, but the Compound Amount is given for every interest rate from 4% to 15% using increments of 0.5%. Use the spreadsheet to determine the minimum interest rate compounded annually that causes $8300 to double in value in 4, 5, and 10 years.

29) Create a spreadsheet in which the Principal, Interest and Time remain the same, but the Compound Amount is given for each of the following compoundings: Annually, Semi-annually, Quarterly, Monthly, Weekly, Daily (365), Hourly (8760), Minutely, Secondly. Consider $1000 deposited for 10 years at 7% interest. Explain your results.

30) Alice is 28 years old and plans to deposit $1000 into a retirement account at the end of each year for the next 30 years. The retirement account is expected to pay about 5.6% per year, compounded annually. Create a spreadsheet which shows Alice's balance at the end of each year for the next 30 years. Change the interest rate to 5.5%; how does the balance change?

31) Out of each monthly paycheck George is depositing $100 into a retirement account which is expected to earn 7.55% per year compounded monthly. Create a spreadsheet which shows George's balance at the end of each year for thirty years. Change the rate to 7.5%; how does the balance change?

32) Create a spreadsheet that keeps track of the amount in a savings account each **year** for 20 years if $50 is deposited at the end of each month and the account pays 9% compounded monthly. Arrange the information under headers:
Deposit   Dep Per Yr   No. Yrs   Rate   Comp Per Yr   Final Amount

---

3   LOANS AND AMORTIZATION

Suppose a bank loans us $10,000. When we pay it back, we not only have to pay back the $10,000 but also the interest that the bank could have earned on that money if it had been invested at the current interest rate. For example, if the current interest rate is 8%, and the bank invested $10,000 for five years then the bank would end up with

$$A = 10,000(1 + 0.08)^5 = \$14,693.28.$$

Now suppose we borrowed $10,000 and agreed to repay the loan in five payments made at the end of each year. Would yearly payments of $2,000 be enough to repay the loan? The bank could earn interest on those payments of $2,000 after they were made. From the payments the bank would end up with:

$$A = \frac{2000[(1 + 0.08)^5 - 1]}{0.08} = \$11,733.20 \ < \ \$14,693.28.$$

This is not enough to repay what the bank could otherwise earn. What if payments of $2,500 were made? The bank would end up with:

$$A = \frac{2500[(1 + 0.08)^5 - 1]}{0.08} = \$14{,}666.50 \ < \ \$14{,}693.28.$$

This is getting closer to the total amount we need to pay the bank, but more trial-and-error is needed. (Banks like their loans to work out to exactly zero!). It would be really nice to have a formula that relates the amount borrowed (principal) to the periodic payment. What were we doing in the example above? We were essentially setting the future value of the **lump sum** borrowed equal to the future value of our **regularly scheduled payments**:

$$P(1 + r)^n = \frac{R[(1 + r)^n - 1]}{r}$$

Multiply both sides by $(1 + r)^{-n}$

$$P = \frac{R(1 + r)^{-n} [(1 + r)^n - 1]}{r}$$

and simplifying

$$P = \frac{R[1 - (1 + r)^{-n}]}{r}.$$

Here is the formula relating the amount of a loan, **P**, to the amount of the payments, **R**, where, as before, **r** is the interest rate per compounding period and **n** is the number of payments. We can solve this equation for **R**, which gives us:

Loan Formula

$$R = \frac{Pr}{\left[1 - (1 + r)^{-n}\right]}$$

Remember that **r** and **n** need to be adjusted by the number of compoundings per year. Let's finish the example we started with:

Example 3.1: We borrow $10,000 from the bank payable annually at 8% for 5 years. What will be the amount of the payments?

Solution: **P** = 10,000, **n** = 5, **r** = 0.08, and we are looking for the monthly payments, **R**.

$$R = \frac{Pr}{[1 - (1 + r)^{-n}]} = \frac{(10000)(0.08)}{[1 - (1 + 0.08)^{-5}]} = \frac{800}{0.3194168} = \$2504.56$$

Example 3.2: Eric buys a house costing $50,000 and makes a down payment of $8,000. He will make monthly payments on the unpaid balance for thirty years and the interest rate is 9.25%. What will his monthly payments be? How much will Eric pay for the house? How much of the total is interest?

Solution: Again we are looking for **R** in the loan formula. The annual interest rate is 0.0925 so the monthly rate is **r** = 0.0925/12 = 0.00770833.

P = $50,000 - $8,000 = $42,000, and **n** = 30x12 = 360.

$$R = \frac{(42,000)(0.00770833)}{[1 - (1 + 0.00770833)^{-360}]}$$

$$R = \frac{323.75}{[1 - 0.063016459]} = \$345.52$$

Since his monthly payments are \$345.52, in one year he will pay 345.52x12 = \$4,146.24. In thirty years he will pay 4,146.24x30 = \$124,387.20. Adding in the \$8,000 down payment, Eric will pay \$132,387.20 for the house, including \$132,387.20 - \$50,000 = \$82,387.20 in interest.

YOU TRY IT 3.1

A small business buys a copier for \$4300. They promise to make quarterly payments with interest of 12% per year for 2 years.
a) What will be the amount of their quarterly payments?
b) How much will they actually have paid for the copier after the 2 years is up?

YOU TRY IT 3.2

Dave pays a $1500 down payment on a car costing $13,899. He finances the rest making monthly payments for the next 3 years at 12.6%.

a) What will his monthly payments be?

b) How much will he pay for the car? How much interest will he pay?

c) If the bank allowed Dave to pay the loan back over one and one half years instead of 3, would you expect his monthly payments to be

- about twice the amount of the payments on the 3 year loan?
- more than twice the amount of the payments of the 3 year loan?
- less than twice the amount of the payments of the 3 year loan?

Make a guess first using your intuition, then compare the actual figures. Also compare the amounts of interest paid in the two scenarios.

Amortization of Loans

To answer questions about what happens if we pay off loans early, we need to examine in more detail how a loan works. Suppose a company buys a set of car phones for its employees. The total bill is $10,000, which it promises to pay off with end-of-year payments over the next 4 years at a rate of 8% interest on the unpaid balance. Using our loan formula, we calculate the payments to be:

$$R = \frac{(10,000)(0.08)}{[1-(1+0.08)^{-4}]} = \$3019.21$$

Consider what happens at the end of the first year. The first **payment** of $3019.21 is made. What portion of the payment is applied to the debt of $10,000 and what part is paying interest? This is the way it is usually calculated:

At the time of the first payment, the company has had the $10,000 for one year. It has promised to pay 8% annual interest on the money owed, and so owes $(0.08)(10,000) = \$800$ in **interest**. This amount is subtracted from the $3019.21 of the first payment and what is left is applied to reduce the debt of $10,000 (i.e., applied to the **principal**.) We summarize the breakdown with the following table:

Payment Number	Amount of Payment	Portion to Interest	Portion to Principal	Balance Due
0	0	0	0	$10,000
1	3019.21	(0.08)(10,000) = 800	3019.21-800 = 2219.21	10,000-2219.21 = 7780.79

The table above is called an **amortization schedule**. A loan is **amortized** if the total principal and interest owed are both reduced every time a payment is made. However, the principal and interest are not reduced by the *same* amount in each payment. Let's add a few more payments to the table to see what happens:

After the first payment the actual debt is only reduced by $2219.21 leaving the company owing $7780.79 at the end of the next year. It is 8% of *this* amount that will be owed as interest.

Payment Number	Amount of Payment	Portion to Interest	Portion to Principal	Balance Due
0	0	0	0	$10,000
1	3019.21	(0.08)(10,000) = 800	3019.21-800 = 2219.21	10,000-2219.21 = 7780.79
2	3019.21	(0.08)(7780.79) = 622.46	3019.21-622.46 = 2396.75	7780.79-2396.75 = 5384.04
3	3019.21	(0.08)(5384.04) = 430.72	3019.21-430.72 = 2588.49	5384.04-2588.49 = 2795.55
4	3019.21	(0.08)(2795.55) = 223.64	3019.21-223.64 = 2795.57	2795.55-2795.57 = -0.02

To compensate for the 2 cents overpaid, the last payment should be $3019.19.

Now we can answer the question, "If we want to pay the loan off early, how much do we owe?" Looking at the table, we can see that the amount owed is the balance at the end of each row. If the company wanted to pay off the debt after the second payment, it should pay $5384.04.

YOU TRY IT 3.3

Create an amortization schedule for a loan of $2000 to be repaid in annual payments at 10% for the next 3 years.

Payment Number	Amount of Payment	Portion to Interest	Portion to Principal	Balance Due
0	0	0	0	$2,000
1				
2				
3				

Example 3.3: Suppose that in the car phone example, the company elects *quarterly payments* for the four years. This is now a loan of $10,000 to be repaid in quarterly payments over four years at 8% on the unpaid balance. What changes from the example above?

|Solution:

• We'll need to re-calculate the payment ($736.50).

• The payments are quarterly for four years, so our Payment Number column will have (4)(4) = 16 payments in it. (This is **n**.)

• The interest in the Portion to Interest column will need to be the **interest per payment period** which is 0.08/4 = 0.02. (This is **r**.)

Let's look at the **first three payments**:

Payment Number	Amount of Payment	Portion to Interest	Portion to Principal	Balance Due
0	0	0	0	$10,000
1	736.50	(0.02)(10,000) = 200	736.50-200 = 536.50	10,000-536.50 = 9463.50
2	736.50	(0.02)(9463.50) = 189.27	736.50-189.27 = 547.23	9463.50-547.23 = 8916.27
3	736.50	(0.02)(8916.27) = 178.33	736.50-178.32 = 558.17	8916.27-558.17 = 8358.10

YOU TRY IT 3.4

Make an amortization schedule for the first three payments of a new car loan, where $14,500 is borrowed at a rate of 12% and monthly payments are made for 4 years.

Payment Number	Amount of Payment	Portion to Interest	Portion to Principal	Balance Due
0	0	0	0	$14,500
1				
2				
3				

This style of amortizing a loan, where interest is due on the balance owed for each payment, is the most common form of loan repayment. However, it is not the only way a lending institution can set up a payment schedule.

Example 3.4: A city wants to borrow money to build a waste disposal plant. They would like a 4 year loan for $297,000. The lending institution agrees to finance the plant at 6% payable annually, under the condition that the city agrees to pay the principle in equal installments with interest due on the balance. What would an amortization schedule for this loan look like?

> Solution: This situation is very interesting because the loan payment formula is no longer useful. We want to determine the payment as the principal owed (always 297,000 / 4) plus the interest due on the previous balance:

Pmt #	Balance	INTEREST	PRINCIPAL	PAYMENT
0	$297,000	0	0	0
1	$222,750	(0.06)(297000) = $17,820	297000 / 4 = $74,250	17820 + 74250 = $92,070
2	$148,500	(0.06)(222750) = $13,365	297000 / 4 = $74,250	13365 + 74250 = $87,615
3	$74,250	(0.06)(148500) = $8,910	297000 / 4 = $74,250	8910 + 74250 = $83,160
4	$0	(0.06)(74250) = $4,455	297000 / 4 = $74,250	4455 + 74250 = $78705

YOU TRY IT 3.5

Make an amortization schedule for a standard version of the loan described in Example 3.4, i.e., borrowing $297,000 at 6% payable annually for four years. Compare the payments made each year, the total amount of interest paid, and early payoff figures in a regular loan with the special loan situation described in the example. If you were trying to be re-elected to the town council, which loan would you want to get?

Payment Number	Amount of Payment	Portion to Interest	Portion to Principal	Balance Due
0	0	0	0	$297,000
1				
2				
3				
4				

There are other variations on loan repayment schedules that are sometimes used, but they are fairly rare. When borrowing money be sure to ask the loan officer about how the interest and principle will be repaid. If the answer is

a fixed payment consisting of interest owed on the balance with the remaining part of the payment applied to the principal,

then we can use the standard amortization table described in this section.

Notice that none of the examples dealt with mortgages; are these loans somehow different? No! They are usually payable monthly for 10 or more years, so the number of payments will be large. We wouldn't want to do such tables by hand. Fortunately, spreadsheets are designed to make building these tables easier. **interest**, and is the subject of the next section.

Exercise Set 2.3

You may assume the standard loan repayment rules apply in the problems below, unless specifically told otherwise.

1) Jimmy buys a house for $232,000. He makes a down payment of $20,000 and finances the balance. How much are his monthly payments if the current interest rate is 8% and it is a fifteen year loan?

2) Dwight buys a $3,000 engagement ring for his fiancee. He agrees to repay the money in one year at 14% interest. How much will his monthly payments be?

3) Terry wants to buy that awesome red car that costs only $23,000. If he puts 10% down and finances the rest for 4 years at 12.4%, what will his monthly payments be? How much will he have paid for the car in the end?

4) Sam and Sue are buying a house that's for sale for $90,000. They put $10,000 down and plan to finance the balance. Compare the pros and cons of the 15 and 30 year mortgages (length of the loan) if their interest rate is 7.5% and they make monthly payments.

5) Nat and Bethany want to buy some land that costs $40,000. They put $4800 down and plan to finance the balance. Compare the pros and cons of a 5 year loan versus a 10 year loan at 9.6% interest, making monthly payments.

6) Linda buys a car for $12,000. She pays $1500 down and finances the rest, paying interest on the unpaid balance. The bank suggests a four year loan at 11.24%. If Linda agrees, how much would her monthly payments be?

7) A local business decides to invest in a new copier costing $3600. They pay $500 down and agree to pay the balance in annual payments over the next three years, paying 14% on the unpaid balance. Make an amortization schedule for this loan.

8) Christa is buying a house costing $54,000. The lowest interest rate is 8.25%. Find her monthly payments if she gets a thirty year loan. What are her payments if she gets a fifteen year loan? How much money will she save with the fifteen year loan?

9) Jerry is buying a car which is priced at $21,225. He puts 10% down and finances the rest at 12%, committing to monthly payments over the next 5 years. Prepare the first four payments of an amortization schedule for Jerry's loan.

10) Ashleigh decided to buy that beautiful 10 acres just off the Blue Ridge Parkway. She can purchase the land for $42,000, and the owner has agreed to allow her to make payments each quarter for the next 5 years at 10% interest on the unpaid balance. Prepare the first four payments of an amortization schedule for the loan. How much money would she need to pay off the loan after the first year?

11) Prepare the first three lines of an amortization schedule for a home loan of $75,000 to be repaid in monthly payments over 30 years at a rate of 8.4% on the unpaid balance.

12) Prepare the first three lines of an amortization schedule for a home loan of $125,000 to be repaid in monthly payments over 25 years at a rate of 7.8%.

13) Tom wants to buy a car. He figures he has $200.00 a month to spend on loan payments. The bank's current new car loans are: 7.25% for 3 years, 7.5% for 4 years and 7.75% for 5 years.

a) Find the amount he can afford to borrow under each loan.

b) If he has $1,500 in cash for a down payment, how expensive a car can he afford under each loan?

14) Write a response to the given memo, justifying your response.

```
To:     J. Student                                    MEMO
From:   R. Stewart, Alumni Association
Re:     Loans

We have the opportunity to purchase a new building costing $245,000.
Please compare our mortgage choices, as listed below:
(1)  Borrow 85% of the cost at 7.5% payable monthly for 15 years
(2)  Borrow 80% of the cost at 7% payable monthly for 20 years
```

15) Write a response to the given memo, justifying your response.

```
To:     J. Student                                    MEMO
From:   R. Stewart, City Planning
Re:     Loans for Waste Treatment Facility
Our town needs to expand the waste treatment facility.  We will need
to finance $370,000.  We have been offered two different types of
loans.  Please compare our choices, as outlined below:
(1)  a standard loan at 9% payable monthly for 10 years
(2)  a custom loan in which we would pay payments on half of the
principal at 9% payable monthly for 10 years, and then pay the other
half of the principal in full at the end of the loan period.
```

The following exercises require access to a spreadsheet.

16) Prepare the entire amortization schedule for the loan described in problem 6.

17) Prepare the entire amortization schedule for the loan described in problem 9.

18) Prepare the entire amortization schedule for the loan described in problem 10.

19) Create an amortization table for a loan of $8900 payable quarterly to be paid off in 4 years at 12%.

20) After the down payment, Sue expects to borrow $12,000 to buy a car. She plans to make monthly payments, paying 13.5% interest. Create a spreadsheet showing her different payments if she repays the loan in 2 years, 3 years, 4 years and 5 years. Include a column which shows how much interest she pays in each case.

21) Ralph and Elaine would like to know what difference a few percentage points will make in their monthly house payment. Create a spreadsheet which shows their payment if they borrow $93,000 for 30 years at rates starting at 5% and increasing 0.5% each time up to 15%. Include a column showing the total interest paid in each case.

22) The Martins plan to buy a house which sells for $89,995. They pay a down payment of 10% and finance the rest at 9.631% interest planning to make monthly payments. Create a spreadsheet in which you compare the changes in Payment, Total Paid for House and Total Interest Paid for loan periods of 10 years, 15 years, 20 years, 25 years and 30 years.

23) Build an amortization schedule for a mortgage to be paid monthly for 30 years at 7.25% if the amount borrowed is $125,000.00. If an extra $100.00 is paid each month on the principal, how many payments must be made? How much interest is saved?

24) Build an amortization schedule to answer the following question: John borrows $12,500.00 at 8% monthly for 4 years to finance a new car. He wants to figure out how much it will take to pay off the loan early. How much will he owe at the end of 2 years? 3 years?

4   PERSONAL FINANCE
-- CREDIT --

Now that you've learned the basics of interest and amortization, it's time to learn how all this financial "know-how" can be used in your personal finances. In the next two sections we will investigate some investment strategies and the realities of financing a major purchase. We'll start with a look at credit. Personal credit comes in many forms. Any financial transaction in which you borrow money, no matter how short the time period, is part of and affects your credit. The most common forms of personal credit are credit cards and loans.

Credit Cards and Interest

Some forms of credit are readily available, particularly open-ended credit in the form of credit cards: Bank cards (e.g., VISA), department store cards, gasoline cards, the list goes on. Once you obtain the card, it is not necessary to apply for credit every time you make a purchase. This type of credit is very convenient. However, such convenience can result in long-term debt if the consumer does not understand the implications of interest.

Charging a purchase on a credit card is equivalent to borrowing money. Most credit cards allow a grace period during which no interest is applied to the amount borrowed. If the balance is paid in full within this specified time period (usually 20-30 days), then no additional interest will be charged to the account. However, if a balance is carried beyond the grace period, then interest is applied, usually at a high rate. But interest rates alone do not determine how much interest you will pay. Consider the following situation:

Jack and Jill had different credit cards, but the rates on the credit cards were the same. They both carried a $250 balance. Yet when they received their bills, Jill's bill was several dollars higher than Jack's.

How can two credit cards that charge the same interest rate end up costing you different amounts in finance charges for the same purchase? Well, there are actually a variety of ways in which credit card companies go about determining the amount of the balance on which you will pay interest each month. Here are three common methods of applying interest, given in descending order of desirability.

Average Daily Balance (ADB) Excluding New Purchases:

In this situation, interest is paid only on any balance left over from the previous month. You get approximately one month's grace period before interest starts accruing on new purchases, regardless of whether you pay your balance in full or not. It is a law in some states that this method be used for all credit cards.

Average Daily Balance (ADB) Including New Purchases:

This method is the most common one available, and can come with or without a grace period. With a grace period, it is very similar to the one stated above, except: If any portion of your balance is carried over from a previous month, there is no grace period for *additional* purchases. Without a grace period, interest is charged from the day of purchase even if you do not carry a balance.

Two-Cycle Average Daily Balance Including New Purchases:

This method can also come with or without a grace period. With a grace period, if you pay your payments in full each month there is no interest charge, and (as is the second method above) you lose your grace period if you carry forward a balance. Furthermore, the grace period you received the first month is retroactively eliminated any time you begin carrying an unpaid balance. Without a grace period, interest is accrued from the date of purchase, and any balance carried forward is charged double interest.

To illustrate the difference between these three methods of computing finance charge, let's look at an example:

Example 4.1: Suppose you make two purchases, one $100 purchase on April 1 and one $200 purchase on May 1. You receive a bill in May for the April purchase, but you do not pay it. However, you pay your June bill in full. The credit card company's billing period ends the end of each month. The annual interest rate is 18% (1.5% a month). Find the finance charges for each method, assuming a grace period.

Solution: Here's what happens:

ADB Excluding New Purchases

Charges from April (May bill):
   $100 charged - grace period is in effect; no interest will be charged.
Charges from April and May (June bill):
   $100 carried over - grace period has expired; interest will be charged.
      Amount of interest: ($100) x (.015) = $1.50
   $200 charged - grace period is in effect; no interest will be charged.
The bill in June will be $100 + $200 + $1.50 = **$301.50**

ADB Including New Purchases (with grace period)

Charges from April (May bill):
   $100 charged - grace period is in effect; no interest will be charged.
Charges from April and May (June bill):
   $100 carried over - grace period has expired; interest will be charged.
      Amount of interest: ($100) x (.015) = $1.50
   $200 charged - no grace period because of balance carried forward; interest will be charged.
      Amount of interest: ($200) x (.015) = $3.00
The bill in June will be $100 + $200 + $1.50 + $3 = **$304.50**

Two-Cycle ADB Including New Purchases (with grace period)

Charges from April (May bill):
   $100 charged - grace period is in effect; no interest will be charged.
Charges from April and May (June bill):
   $100 carried over - grace period previously allowed is now retroactively eliminated. Interest will be charged on the $100 for the previous month:
      Amount of interest: ($100) x (.015) = $1.50
      For the second month, interest will be charged on the first month's balance, which now includes principal *and* interest:
      Amount of interest: ($101.50) x (.015) = $1.52
   $200 charged - no grace period because of balance carried forward; interest will be charged.
      Amount of interest: ($200) x (.015) = $3.00
The bill in June will be $100 + $200 + $1.50 + $1.52 + $3 = **$306.02**

The same interest rate can result in $1.50, $4.50, or $6.02 in interest, depending on the method of calculation. These three are just a few of the methods used. Obviously, how a company applies an interest rate is an important factor to be aware of when considering the relative merits of different credit cards.

So far, we've looked at interest rates as being fixed. Unfortunately, credit card rates are often **variable rates**, meaning that the interest rate can change every billing period! The most common way of varying rates is to tie the credit card rate to the highest Prime Rate for the billing period as reported in *The Wall Street Journal*. For example, the rate may be set at 8% plus Prime; if the highest Prime rate that month was 9%, you'd be paying 17% interest!

In addition to charging interest on unpaid balances, many credit companies charge an **annual fee** for the use of the card. This fee can go as high as $250 (for an American Express Platinum Card), but the normal range is $20 to $60 per year.

Creditors make money when interest is charged, so they try to discourage the cardholder from paying the balance in full. One way to do this is to set a **minimum payment**. The amount of the minimum payment is dependent on the current balance of the account, but in most cases the minimum payment is very low. Suppose you charge $195. When your bill comes, the amount owed will be $195 (assuming a grace period). There may also be a minimum payment option as low as $10. By making the minimum payment, the creditors can charge interest on the $185 balance next month! If the borrower regularly pays only the minimum payment, the credit card company will do quite well.

Example 4.6: John makes a $50 charge on his credit card. His credit card rate is 19.5%, or 1.625% monthly. $10 is the minimum payment required, and John chooses to pay this each month. What will his total interest be?

Solution: Let's look at an amortization table for John's loan payments:

Bill #	Payment Amt	Interest	Principal	Balance Due
1	0	0	0	$50
2	$10	$.81	$9.19	$40.81
3	$10	$.66	$9.34	$31.47
4	$10	$.51	$9.49	$21.98
5	$10	$.36	$9.64	$12.34
6	$10	$.20	$9.80	$2.54
7	$2.58	$.04	$2.54	-0-

Total Interest: **$2.58**

If John had paid the remaining balance in full after the second bill, then he would have paid $0.81 in interest. By making the minimum payment every time, he pays $2.58 in interest. If he had made additional charges along the way, the situation would have been even worse.

The minimum payment scenario can be complicated by a "minimum finance charge". Most companies have a minimum amount that you must pay if you incur a finance charge. Even if the calculated finance charge is $0.04, you might have to pay as much as $1.50 instead. Let's consider Example 4.6 again.

Example 4.2: Suppose that John's credit card has a minimum finance charge of $0.50. What will his total interest be in this case?

Solution: The payments John makes for Bills 5, 6, and 7 change the table:

Bill #	Payment Amt	Interest	Principal	Balance Due
1	0	0	0	$50
2	$10	$.81	$9.19	$40.81
3	$10	$.66	$9.34	$31.47
4	$10	$.51	$9.49	$21.98
5	$10	$.50	$9.50	$12.48
6	$10	$.50	$9.50	$2.98
7	$3.48	$.50	$2.98	-0-

Total Interest: **$3.48**

Many credit card companies have now expanded their services to include **cash advances**, where you can actually withdraw cash and charge it to your credit card. It is common practice for interest charges to begin immediately, i.e., there *is no grace period for cash advances made on your credit card!* Many banks also charge higher interest and a transaction fee for this service. The convenience of the cash can be costly.

When applying for a credit card, you need to consider:

- interest rate (variable?)   - grace period
- method of charging interest   - annual fee
- minimum interest payments   - fees for cash advances

This information must be disclosed somewhere on the application. Most credit cards have adopted a standard summary table similar the one below for displaying this information.

Annual Percentage Rates	Currently 12% for purchases; 19.8% for cash advances
Variable Rate Information	The annual percentage rate for purchases may vary each calendar quarter. We will calculate the variable rate by adding 9.4% to the rate disclosed as the US Prime Rate in the Wall Street Journal on the third Tuesday of each quarter. This rate will not be lower than 11% or higher than 19%
Grace Period for Payment of Balances for Purchases	20-25 days
Method of Computing the Balance for Purchases	Two-cycle average daily balance (including new purchases).
Annual Fees	None
Minimum Finance Charge	$1.00
Transaction Fee for Cash Advances	$1.75
Late Payment Fee	$15
Over -the Credit-Line Fee	$10

Personal Credit

One last word about credit: Each person who has a credit card also has a credit rating, whether or not they are aware of it. Paying off a credit card can take on many different forms, anywhere from paying the total balance each month (the *best* use of credit) to paying a monthly minimum payment (the *worst* use of credit). But if you don't pay regularly, you will end up with a bad credit rating that can follow you around for seven years or more. One local bank reported that any payments that are 30 days late are automatically reported to the Credit Bureau by the computer -- no questions asked. If your credit rating is bad, your efforts to borrow money to finance large purchases (such as cars or houses) may be met with refusals by lending institutions. Even employers will sometimes consider credit ratings of prospective employees when making hiring decisions. The abuse of credit is widespread and costly, but an informed consumer can make good choices for how to use credit appropriately.

If you have a question about your credit rating, contact your local credit bureau. For around $15, you can get a copy of your report to check for mistakes. You may be surprised by what you find; even late phone bills can affect your credit rating. If you do have trouble making payments on a loan, talk to a banker. Often he or she can work out a repayment scheme that you can afford and that will preserve your good credit rating.

What Does It Really Take To Finance A Major Purchase?

Buying a new car is the first major purchase for many young people. If college students are asked what they look forward to after finding a job, they will often begin to describe their bright red sports car with the incredible sound system. But there are a lot of factors that come into play when you consider buying a car. For instance, where do you start?

When several bankers and car dealers were asked what their main advice would be to first time car buyers, it was: Only buy what you can afford. If you can't make the payments, you may lose the car, lose the money you spent up to that point, and ruin your credit for 7 years! It is certainly good advice. You can determine what you can afford by looking at your *debt to income ratio*.

Debt-to-Income Ratio

When you visit a bank or other lending institution you will be asked to fill out a loan application. The purpose of the loan application is to determine if you are a good risk. The application requires information about your:

- credit history: Have you been consistent in paying your bills, loans and other obligations over the past several years? Have you frequently overdrawn checking accounts or been delinquent in making payments?

- stability: How long have you been at your present job and in your present living situation? Is your income likely to change soon?

- current income and assets: What is your annual income? How much do you have in your bank accounts? Do you have any other assets of property (cars, real estate, etc.) or investments (stocks, bonds, etc.)?

- current debt: How much do you owe today, and to whom? How much might you owe tomorrow if you used your credit card to make multiple purchases?

After doing a thorough check on the information you have supplied (the bank verifies your numbers by contacting a licensed credit agency), the bank determines if you are a good risk. To see if you can afford the amount you've requested, the bank considers your total monthly debts including your expected loan payment in light of your total monthly income. Most banks decide if you can afford the loan by calculating your monthly **debt-to-income ratio**:

$$\text{Debt-to-Income Ratio} = \frac{\text{total debt}}{\text{total income}}.$$

If your expected debt-to-income ratio is higher than the bank's cut off (most banks have a limit between 35% and 40%), your loan will most likely be denied.

Definitions of "debt" vary by the institution. For instance, debt may or may not include monthly bills for rent/utilities/phone. Generally the percentage is adjusted higher if these are included. The individual institution will clarify their methods when requested.

Example 4.3: Robert just graduated from college and has begun a new job. He promised himself that he would buy a new car. Robert calls his bank and they give him the following information:

- Loans require a debt-to-income ratio of 40% or less.
- Debt includes all outstanding debt (current credit card debt, loans) as well as regularly occurring bills (rent/utilities/phone, insurance)
- Income is gross income, before any taxes or other reductions.

Here is Robert's information. His job pays $22,800 per year (gross). He currently has $400 in credit card debt with a minimum payment of $35 per month, and a $75 per month student loan payment. His apartment costs $325 per month; his electric bill runs about $22 per month and his phone bill about $26 per month. Robert's employer pays his health insurance, but his car insurance is about $360 per year.

What is Robert's debt to income ratio now, before he borrows to buy a car? What is the highest payment he can afford without going over the maximum ratio?

Solution: Income (monthly): $22,800 ÷ 12 months = $1900

Debt (monthly):

Credit Card:	$35
Student Loan:	$75
Rent:	$325
Electric:	$22
Phone:	$26
Insurance:	$30 (360 ÷ 12)
Total:	$513

Debt ÷ Income: 513 ÷ 1900 = 0.27

27% debt to income ratio

Maximum payment can not exceed 40%: $\dfrac{\text{debt}}{\$1900} = 0.40$

Multiply both sides by $1900 to get maximum debt = $760

Since Robert's current debt is $513, the maximum payment he can afford without exceeding the 40% ratio is $760 - $513 = $247

YOU TRY IT 4.1

Linda wants to buy a new car. She goes to her bank and finds out that the maximum debt-to-income ratio is 40%, and for the formula, debt includes major expenses like loan payments, housing and insurance. Linda rents an apartment at $375.00 per month. She is paying off a bill at Fabulous Furniture at a rate of $22.50 per month. Her car insurance is $210 every six months, and she expects it to increase to about $300 every six months for the new car. Linda just began her second year teaching at Happy High School, where her gross income is $20,400. She works in the summer at a pharmacy and usually earns an additional $3000. What is Linda's debt to income ratio now? What size car payment could she afford?

Down Payments

Even if your credit is excellent, the bank may require you to invest up front in the car. Many banks have a minimum down payment percentage; usually it is stated, "We will finance up to 90% of the selling price for a 60-month loan." (This percentage will vary with the institution, the value of the car and the term of the loan.) This means that you must provide at least 10% of the selling price of the car before the loan can be approved. The down payment may be in the form of cash or trade-in value of your older car.

Example 4.4: Robert (from Example 4.3) has little cash savings, but he is planning to use the trade-in value of his old car as his down payment. His bank advises him that his car has a trade-in value of about $2800. If his bank will finance 90% of a new car purchase at 9% for 36 months, what's the most he could spend on the new car?

Solution: $2800 is 10% of what number? $2800 = 0.10 \times x$
$$x = 28,000$$

So $28,000 is the most expensive car for which he has a down payment.

Can he afford $28,000 based upon the debt-to-income ratio information computed in Example 4.8?

We found that his maximum payment is $247.00. To see how much he can borrow total, we'll need the loan formula, where: R = 247, n = 36 and r = 0.09/12 = .0075

$$247 = \frac{P\,(.0075)}{1 - (1+.0075)^{-36}}$$
$$P = \$7767.36$$

Adding the $2800 down payment, the new car should sell for less than
$$\$7767 + \$2800 = \mathbf{\$10{,}567.}$$

Clearly, he can't afford a $28,000 car. The bank would refuse to lend him that much money.

YOU TRY IT 4.2

Calculate Robert's maximum car price in Example 4.9 if the loan is for 48 months at 9%; 60 months at 9%.

Potential Increases To Payment

When you actually apply for the loan at the bank, you may find that your payment ends up higher than your estimate. Many banks have a one time "Initial Transaction Fee" or a "Pre-Paid Finance Charge" of $50 or more just for processing the loan. Usually this is added on to the amount borrowed, although it may be paid separately.

Because the future is unknown, most lenders also offer to finance two types of loan insurance: Life and Disability Insurance. Life Insurance would pay off the loan in case of the death of the debtor. Disability Insurance would make car payments for you if you became unable to work for some period of time. Buying these insurance policies can add from $10 to $100 to your monthly payment, depending on the type of coverage and age of the applicant. A little investigation may yield substantial savings here. Usually a simple Term Life Insurance policy may be purchased separately from an insurance company at better rates, especially for young people. Often disability insurance is provided through your employer. Although it is certainly convenient to just add these costs to your payment, you may be paying substantially more for the service.

Also required, and varying from state to state, are sales tax, title fees and proof of insurance. These fees can add a significant amount to the purchase price, so be sure to ask about them.

Example 4.5: Suppose Robert (Examples 4.8 and 4.9) decides to add "credit insurance" in the form of life and disability insurance to his regular payments. The bank estimates that it will cost him an additional $32 per month. If he wants to keep his payment at $247 per month (using the 36 month loan term), how will his maximum price for the car be affected?

Solution: Subtracting the $32 from his $247 available for a payment
$$\$247 - \$32 = \$215$$
Now recalculate the top loan amount with this payment:
$$215 = \frac{P\,(.0075)}{1 - (1+.0075)^{-36}}$$
solving, $\qquad P = \$6761.06$
Adding the $2800 down payment, the new car should sell for less than
$$\$6760 + \$2800 = \mathbf{\$9560}$$
This is a considerable difference, so Robert considers a longer term loan.

YOU TRY IT 4.3

Calculate the change in his car price if Robert chooses only the life insurance for $12.50 per month in Example 4.5.

What About Leasing A Car?

In the 1990's, the average price of a new car rose to almost $20,000. Due to these higher prices and resulting higher down payment requirements, new car leasing has become much more attractive, especially for someone who wants to drive a luxury car but cannot afford to purchase one. Initial outlays and monthly payments may be lower, so people that don't have the down payment needed to buy a new car might choose a lease.

Perhaps the best way to explain leasing is as only paying for the value of the car's life that we use. For example, we lease a $12,600 Chevrolet Cavalier for 3 years. The dealer estimates that the value of the car after that time will be 43% of its original value. This is termed the "residual value" of the car.

$$0.43 \times 12,600 = \$5,418$$

We make monthly rental payments sufficient to pay the amount still owed on the loan down to the residual value ($5,418 in this case), instead down to 0. So we need to build a new formula. Remember how we built the loan formula? We set the lump sum formula equal to the regular payment formula because we wanted to figure what payment would come out the same as if the bank had invested the principal instead of giving it to us. We want to use a similar idea, except now we want the payments *plus the residual* to come out the same as the lump sum. Let's call the residual E (we've got too many things called R already!):

$$\frac{R\left[(1+r)^n - 1\right]}{r} + E = P(1+r)^n$$

Hence to figure the payment R, we substitute P = original price, r = interest on lease, n = length of lease, and E = residual value. If our rate is 10.5% payable monthly for a 36 month lease:

$$\frac{R\left[\left(1+.\frac{.105}{12}\right)^{36} - 1\right]}{\frac{.105}{12}} + 5418 = 12600\left(1+\frac{.105}{12}\right)^{36}$$

Solving this for the payment R gives (Try it!):

$$R = \$280.84.$$

Another way to do this if we have access to an amortization table on a computer, is to experiment with the payment until the balance at the end of the lease (in this case the 36th month) is equal to the residual value of the car. Trial and error can quickly arrive at the lease payment this way.

What else do we need to look for in a lease? Lease **deposits** vary from car to car; the dealer tells us that ours is $325 plus the first month's payment; the $325 is refunded when the car is turned in. At the end of 3 years we return the car to the dealer or pay the residual value. Below is a comparison:

New Car Cost $12,600
Rate   10.5%
36 month loan versus lease

PURCHASE   **LEASE**

PURCHASE		LEASE	
Down payment (20% min)	$2520	Deposit	$325 + first month's pmt
Monthly Payment	$327.62	Monthly Payment	$280.84
Trade-in or Sell for	$5418	Return Car to Lessor and get back $325	

In this instance, the lease option may very well be the best buy for someone. Of course, a well maintained car might be worth more than $5,418 after three years, so it's possible that the trade-in value will be higher. Many lease options are available and could be viable choices for first time buyers, but the details of the fees associated with the lease will vary from car to car and dealer to dealer. The main message from bankers, car dealers, and consumer publications is:

Read the fine print!

There are often a variety of charges which apply at the beginning or end of the agreement. The lending institution also may require the last month's payment in advance, in addition to the deposit. There may be transaction fees, termination fees, etc. Some leases require costly low deductibles on your car's insurance. If you drive more than a certain number of miles a year, you could owe as much as 10¢ to 25¢ per mile for the excess mileage when you return the car. (Suppose the number of allowable miles per year was 15,000 and the fee for additional mileage was 25¢; if you drove an extra 5,000 miles a year, you would owe an $1250!) If you carry equipment or small children, you may face high "excessive wear" fees.

In spite of these possible additional charges, leasing is becoming an increasingly better option as we see from the Chevy Cavalier example. Leasing a car that holds its value can be very cost effective. Consumer magazines can be extremely helpful in keeping readers informed about pitfalls in car buying and leasing. Such publications are available in most public libraries.

Example 4.11: As Robert was driving to the dealership to look at new cars, his engine blew irreparably. Suppose the car dealer will lease a car for 36 months at a loan rate of 10.8%. He will calculate the residual value at that time to be 42% of the new car price. If Robert wants to stick with his $247 payment (no extra insurance), what is the highest price car he can afford to lease? Assume he can make the required initial deposit and fees out of his savings and/or remaining car value.

> Solution: Robert basically needs to make loan payments of $247 for 3 years at 10.8%. The maximum value of the car he could use would be:
>
> $$\frac{247\left[\left(1+\frac{.108}{12}\right)^{36}-1\right]}{\frac{.108}{12}} + 0.42P = P\left(1+\frac{.108}{12}\right)^{36}$$
>
> solving, $P = \$10{,}874.56$
>
> So, the new car price would have to be no more than $10,874.56.

What Changes when Buying a House?

When a home is purchased, the loan is called a mortgage. Most of the requirements we described in our discussion of car loans still hold:

Banks will look at the borrower's **debt-to-income ratio** for mortgages as well. There is usually an additional rule-of-thumb imposed: *The monthly mortgage payment shouldn't exceed 28% of the monthly income.* The bank's usual ceiling (often around 40% as mentioned before) is also imposed on the total debt.

Banks will require **down payments**, usually at least 20%. Borrowers paying less than 20% may be required to purchase *private mortgage insurance* (PMI) to protect the lender's investment. This is usually added to the monthly loan payment. Few people realize that after paying off part of the mortgage they can have the PMI canceled. Since PMI only protects the lender, it is best avoid it completely or cancel it as soon as possible.

Another **addition to the monthly payment** not usually found in car loans is the *escrow* payment. The escrow account for a mortgage is set up to accrue the annual property tax and property insurance for the house. There are advantages and disadvantages to this. On the positive side, making this payment every month means that when tax and insurance payment time comes around, the money is ready to be sent right from the bank to the local tax collector and insurance agent. The disadvantage is that the bank keeps any interest earned on the money in the escrow account, which could be a sizable amount each year if the taxes and insurance are high. Some banks will waive the escrow account, usually for a fee. If the borrower is good at remembering to save up for tax and insurance payments, the interest earned on the money can make up for any initial fees in a few years.

Transaction fees are more commonly called *closing costs* on mortgages, and are much higher than for car loans. These include an origination fee for the lender, appraisal fee for the appraiser, title search and recording fees for the attorney, and points. Points are one-time lump sum fees charged by the lender and are equivalent to 1% of the principal for the mortgage. Points are essentially up-front interest charges, so banks that charge several points tend to have lower interest rates. This can make mortgages hard to compare. On the plus side, interest on a home is tax deductible, and since points are essentially interest, they are almost always tax deductible as well. Here is an example to illustrate the key ideas.

Example 4.12: John has found a house to buy and approaches a bank in search of a mortgage. He needs to borrow $100,000 and has passed all of the credit checks. The bank offers him a 30 year loan at 8%, with the option of buying up to 2 points, each of which drops the interest rate by 1/4%. What should he do?

Solution: John's house payments when buying 0, 1 and 2 points are:

0 points: $\dfrac{\left(\dfrac{.08}{12}\right)100,000}{\left[1-\left(1+\dfrac{.08}{12}\right)^{-360}\right]} = 733.76$ -- costing \$0.00 additional up front

1 point: $\dfrac{\left(\dfrac{.0775}{12}\right)100,000}{\left[1-\left(1+\dfrac{.0775}{12}\right)^{-360}\right]} = 716.41$ -- costing \$1000 additional

$$2 \text{ points: } \frac{\left(\frac{.075}{12}\right)100,000}{\left[1-\left(1+\frac{.075}{12}\right)^{-360}\right]} = 699.21 \text{ -- costing } \$2000 \text{ additional}$$

So what should John do? Here are some things to think about:

Paying two points costs $2000 and drops the payment by $34.55.
How long will it take John to recoup, assuming that he can earn 5% interest monthly on the savings from the payment? We need to find n in:

$$2000 = \frac{34.55\left[\left(1+\frac{.05}{12}\right)^{n}-1\right]}{\frac{.05}{12}}$$

Solving this gives n = 52 months or 4 years and 4 months. The rest of the interest on the payment savings (360 - 52 = 308 months worth) would be added savings for John. This is, of course, provided that John actually invests the savings and can get an average of 5% interest.

The other thing to keep in mind is that John will have an extra $2000 deduction on his taxes for the year -- provided he pays cash for them up front (i.e., doesn't add them on to his loan amount).

Remember that if John doesn't have the $2000 in cash in addition to his down payment, he can't afford the points in the first place!

Not all Loans Have Fixed Interest Rates

All of the discussion so far on borrowing money has assumed that the interest rate doesn't change over the life of the loan. Most loans, especially car loans, are variable rate loans. This means that the lending institution has the right to change the interest rate according to some leading economic indicator (called the *index*) such as the prime rate. There are legal restrictions on this: The lender must inform the borrower of the size and frequency of such changes (called the *interval*), and the ceiling (called the *cap*) for the rate. For example, a two year *adjustable rate mortgage* (ARM) might have the following particulars:

- 5.75% to start indexed on the 6-month US Treasury Bill
- adjustments of at most 2% are allowed every 2 years
- the cap is 12.5%

We can get around this by thinking in terms of an average rate over the life of the loan, but this is not good enough if we want accurate dollar figures for planning purposes. How does the loan process change with variable rate interest? The initial calculations – down payments, payment amount, etc. – don't change. In fact, in the 2 year ARM above, the first two years of the amortization table don't change. How can we predict future finances with an adjustable rate? One approach would be to assume the worst case (that every time the interest rate can be increased it is by the maximum amount), and make changes to the initial amortization table at the appropriate rows. Then if the rate doesn't climb, or even drops (yes, that sometimes happens!) the borrower comes out ahead of the plan.

One important question to ask the lender when considering an adjustable rate loan is what happens to the payment when the interest changes? Some banks redo the calculations at that point, almost as if the loan were refinanced at the new rate, so a new rate would mean a new payment amount. Others don't change the payment, but instead shorten or lengthen the life of the loan accordingly. Find out in advance what your lender's policy is so that you are not unpleasantly surprised!

Exercise Set 2.4

1) What is the debt to income ratio for an individual who earns $28,500 annually and has the following debts? Debts: Rent at $395 per month, car insurance at $430 every six months, school loan at $145 each quarter.

2) What is the debt to income ratio for an individual who earns $59,200 annually and has the following debts? Debts: Home mortgage at $695.22 per month, car insurance at $520 every six months, child care at $720 each quarter.

3) Sally has been working for several years at a print shop and now makes $8.50 per hour. She works 40 hours per week, 52 weeks per year. She also teaches at a racquetball club some nights and weekends, averaging $200 more income each month. Since Sally rents an upstairs room in a home, her $210 per month rent includes water/electricity/heat. She has her own phone line, which costs her on average $22 per month. Her car insurance is $216 every six months. She has an Exxon card which she doesn't use for gasoline, but just for necessary car repairs. She currently owes $150 on the card and they require her to pay $21 per month. The only other regular bill she has is from the local department store where she has charged clothes and other items. It seems like the account never gets paid off, but she always sends the minimum payment of $35 each month.

a) What is Sally's debt to income ratio?

b) What is the most she could afford for a car payment if her bank considers the limit of this ratio to be 38%?

4) John has the income and debts listed below, and wants to buy a home costing $129,000. He has 20% to put down and has found a bank offering a 30 year fixed rate mortgage at 9.2% payable monthly. If the bank has a mortgage ceiling of 28% and a total debt ceiling of 40%, can John afford this house?

John's Monthly Income and Debts

Income	Utilities	Insurance	Student Loan	Furniture Loan	Car Payment
$2125	$137	$110	$45	$60	$229

5) Priscilla has the income and debts listed below, and wants to buy a home costing $115,500. She has 20% to put down and has found a bank offering a 30 year fixed rate mortgage at 8.2% payable monthly. If the bank has a mortgage ceiling of 28% and a total debt ceiling of 40%, can Priscilla afford this house?

Priscilla's Monthly Income and Debts

Income	Utilities	Insurance	Student Loan	Car Payment
$4400	$155	$160	$163	$429

6) Sally (from Exercise 3) just heard that an apartment across town which she has always liked is coming vacant soon. It is in a great location, close to work, friends and a park. The apartment managers have told her that the rent for a one bedroom apartment is $312 per month; and that she should expect an average of $40 additional for utilities and services. She knows her phone will be about the same, but Cable TV will add $12 per month. If she moves, how will it affect her debt to income ratio and maximum affordable car payment?

7) Jane's car is paid for, but it is 5 years old. When it was new, it sold for $12,500. She has taken good care of the car and it is in good condition. The car has 48,000 miles on it. Checking with a consumer bureau, Jane finds her model car loses about 25% of it's value each year. This applies if the car is in good condition and has no more than 10,000 miles per year of age. About how much is her car worth now?

8) Jane goes to a used car dealer to ask about selling her five-year-old car. When it was new, it sold for $12,500. The dealer told her that the car depreciated at a rate of 33% per year. What is Jane's car worth now?

9) If Sally (from Exercise 3) calls her bank and finds out that the new car loan rate is 12% on a 48 month term, what is the most expensive car she should consider without going over the debt to income ceiling of 38%? Assume that she can trade her car in for a down payment value of $2500.

10) If Sally (from Exercise 6) decides to take the new apartment and finds out that the new car loan rate is 12% on a 48 month term. What is the most expensive car she should consider without going over the debt to income ceiling of 38%? Assume that she can trade her car in for a down payment value of $2500. Should she take the new apartment and still plan to buy the car?

11) David has just started a new job at an annual salary of $23,250 per year. He lives in a nice apartment which costs him $324 a month plus utilities and services of about $48 per month. Phone, TV and cleaning service cost him an additional $74 per month on average. He is repaying a school loan at a minimum rate of $50 per month. His car insurance runs $80 per month (he likes to speed).

a) What is David's debt to income ratio?

b) What is the maximum he could afford for a car payment if his bank considers the ceiling to be 40% on this ratio?

c) If his old car is his only down payment (worth $1200 as a trade-in) and the bank wants at least 10% down payment, what is the most expensive car David could afford? Assume 12% financing for 48 months.

12) David (Problem 11) just heard that he might be able to lease a car. The dealer assures him that he can give his old car for deposit and fees and just start making payments.

a) What price car could he afford on a 36 month lease? Assume a rate of 12%.

b) If the residual value is 40% of the price of the car, what is the maximum price car he could lease?

c) If the cheapest car that the dealer will lease is priced at $12,000, what are David's options?

13) Alice charges $150 on her VISA on January 1. She receives a bill for this purchase in February, but does not pay it. On February 1, she makes another charge of $400. She receives a bill in March and pays it in full. Compute the amount of the March bill for each of the three interest accrual methods discussed in the text. Assume a rate of 18% (1.5% monthly).

14) Automobile lease agreements estimate the value of a vehicle at the end of the lease agreement. This enters into the cost of the lease and is what you can buy the car for at the end of the lease. These values are expressed as a percentage of the original value. Here is a list of several cars available currently through GMAC leases, and their residual percentages at the end of two years. What will the actual dollar residuals be? What amount will you have to finance for a lease?

CAR	LIST PRICE	RESIDUAL
Cavalier	$12,600	53%
Camaro	$18,000	58%
Corvette	$45,000	62%
Astro Van	$24,000	58%
Tahoe	$27,000	70%

15) Don has found a home costing $99,990 and is interested in financing it with a mortgage from the local credit union. They have several mortgage options:

1. A fixed rate 30 year mortgage at 7.5% payable monthly
2. A fixed rate 15 year mortgage at 7.25% payable monthly
3. A 5 year ARM at 6.4% payable monthly in which the interest may be raised or lowered no more than 2% every five years -- with a cap of 12%.

The credit union requires 20% down to avoid PMI and allows the borrower to buy as many as 2 points up front, each reducing the interest rate by 1/4%. Calculate the monthly payments and total paid over the life of the loan in each case, and discuss the pros and cons of the three options. (Assume worst case on the ARM.)

16) Jane has found a home costing $129,950 and is interested in financing it with a mortgage from the local credit union. They have several mortgage options:

1. A fixed rate 30 year mortgage at 8.5% payable monthly
2. A fixed rate 15 year mortgage at 8.25% payable monthly
3. A 3 year ARM at 6.1% payable monthly in which the interest may be raised or lowered no more than 3% every five years -- with a cap of 14%.

The credit union requires 20% down to avoid PMI and allows the borrower to buy as many as 2 points up front, each reducing the interest rate by 1/2%. Calculate the monthly payments and total paid over the life of the loan in each case, and discuss the pros and cons of the three options. (Assume worst case on the ARM.)

The following exercises require access to a spreadsheet.

17) Ed likes to keep track of the balance in his IRA.

a) If he deposits $2000 at the beginning of each year and the account pays 7.5% annually, create a spreadsheet that will show his balance at the end of each year for 30 years.

b) Although Ed has the best of intentions, he only manages to deposit the $2000 for 5 years and then stops. Create a new spreadsheet showing his balance at the end of each year for the 30 years.

18) Create a spreadsheet which shows the annual effective rate for 8% at the various compoundings: annual, semiannual, quarterly, monthly, daily (365).

19) Jeff charges a $1100 computer to his credit card. The interest rate is currently 14.9% ADB including new purchases payable monthly, and the credit card has the following scale for determining minimum payments:

For balances over $1000, the minimum payment is $40 per month.

For balances $501-$1000, the minimum payment is $30 per month.

For balances $251-$500, the minimum payment is $20 per month.

For balances less than $250, the minimum payment is $10 per month.

Assuming that Jeff pays the minimum payment each month, create a spread sheet that displays Jeff's repayment schedule. How long does it take him to pay off the loan this way? How much interest does he end up paying?

20) Alfred charged $1000 on his MasterCard on January 1. The annual rate on his card is 16.5%, and interest is computed using the Average Daily Balance Excluding New Purchases method. Alfred commits to paying $100 a month until the balance is paid off. Alfred is also committed to a regular savings plan. Each month he sets aside $100 into an account paying 5.25%, compounded monthly.

a) Create a spreadsheet showing the amortization schedule for Alfred's credit card debt. How many payments does he have to make? How much interest does Alfred end up paying?

b) How much money will Alfred earn in his savings account (in interest) during this same time period? Considering the amount of interest earned through savings and the amount of interest paid through credit, what was his net loss/gain over this time period?

c) What if Alfred took the $100 he was saving each month and added that to his monthly payment on his credit card? Create a new spreadsheet showing the amortization schedule for Alfred's credit card debt now. How many payments does he have to make? How much interest does Alfred end up paying? Assuming that he picks up the monthly savings payments as soon as the debt is paid, what would be his net gain/loss (over the same time period considered in part (b))?

d) Discuss the merits of paying off debt vs. investing in savings. What is it that makes the difference?

21) Let's compare the costs of buying and leasing a car. You have the choice of leasing or buying a car with a selling price of $15,000 and a list price of

$15,500. At the end of three years it is estimated that the car will have a residual value of 43% of the list price. Interest rates for leasing and buying are both 10% payable monthly.

a) What is the residual value of the car in dollars?

b) Compute the amount that must be financed for a three year lease on the selling price of the car. (Remember to consider the residual value.)

c) Compute the monthly payments for a five year loan to purchase the car. We will assume no down payments are involved, although usually both loans and leases require some money down.

d) Build the amortization table for the loan. How does the estimated residual value of the car compare to the amount left to pay on the loan at the end of three years?

e) Using the numbers you have computed, write a paragraph explaining which choice you would make: Lease or Buy. Argue for your decision, but also mention why the other choice might be good for someone else.

22) Build the three amortization tables for the loan options given in problem 16.

a) Consider the consequences of putting 15% down instead of 20% and paying PMI insurance calculated as 4% of the monthly payment.

b) Add to your work in (a) the escrow account, assuming taxes and insurance work out to be roughly $1500/year.

c) Add to your work in (b) the option of making extra payment towards the principal each month. What happens with $100? $200?

---

5   PERSONAL FINANCE
-- INVESTMENTS --

Investing money to save for retirement, college, down payments on houses, etc., is a situation we all encounter sooner or later. There are so many options that most people consult a financial planner, trusting the planner to make wise decisions on their behalf. A little knowledge of vocabulary and the formulas from this chapter can help you take a more active role in planning your investments. Here are four terms that apply to many different financial instruments:

Principal

As before, the principal is the initial amount invested. This could be a lump sum initial deposit as in a CD or take the form of payments as in an IRA. This can also be called **capital** in certain lump sum situations.

Yield

The yield is the earnings on an investment; this may be stated as a fixed sum or as an interest rate. This is also called the **return**.

Insured versus Uninsured

Most of the investments available from financial institutions like banks fall under *Federal Deposit Insurance Corporation* (FDIC) protection; the FDIC protects up to $100,000 per depositor per financial institution in a savings or checking account, certificate of deposit, money market, IRA or Keogh account. Other investments that are not deposits, such as mutual funds, stocks and bonds are not insured. However, these instruments are regulated by the *Securities and Exchange Commission* (SEC) to ensure fair operation and full disclosure of information. The SEC cannot not guarantee an investor will not lose money.

Tax Deferred versus Tax Exempt

Investments that are tax deferred are only taxed when the investment is liquidated (cashed in). Investments that are tax exempt are never subject to taxation on the interest earned and in a few cases on the principal invested.

Individual Retirement Accounts

An Individual Retirement Account (IRA) is an attractive option for many qualified working people. Its name comes from the fact that money is put into an interest bearing account by a working individual and made available to that individual upon retirement. The reason this is different from simply putting some earnings in a savings account is that any money deposited in an IRA is tax deferred. You pay income tax on the principal and the interest as you draw it out in retirement. At that time it is anticipated that your income will be less, and thus your tax rate lower, so you'll pay less tax on the money. There are income restrictions, but some people can deposit up to $2000 per year in an IRA without paying tax on the principal, making it an attractive *tax shelter* (i.e., you can shelter your money from being taxed). IRAs can be opened at any bank and are protected up to $100,000 by the FDIC.

People who earn too much to get a tax deduction from a traditional IRA might be able to take advantage of a 401K plan through their employers. These plans are similar to IRAs in that they are tax deferred until retirement, but the income limits are much higher and all transactions are handled through employers. These plans are called Keogh plans for self-employed individuals.

Recently a new form of IRA has emerged, called the Roth IRA. In this non-traditional IRA, taxes are paid up front on the principal (just as you pay taxes on money deposited into your checking account), but upon retirement the interest earned is tax exempt. When choosing between a traditional and Roth IRA, many factors must be considered, such as age, probable tax bracket at retirement, present tax burden, etc.

Whether a traditional or Roth IRA, interest is earned in exactly the way discussed in this chapter. Here are two examples to illustrate.

Example 5.1: At the age of 20, Tom deposits $1800 in an IRA that will provide 8% interest compounded quarterly. How much will that one deposit be worth when he begins to withdraw it at age 65?

> Solution: In other words, Tom is making a lump sum deposit of $1800 in an account paying 8% interest compounded quarterly for 65-20 = 45 years. $\mathbf{P}$ = 1800, $\mathbf{r}$ = 0.08/4 = 0.02, and $\mathbf{n}$ = 4 x 45 = 180, so:
>
> $$A = 1800(1 + 0.02)^{180} = 1800 (35.320831) = \$63{,}577.50.$$

If Tom pays 25% of his salary in federal and state income taxes, then he saves $(0.25)(1800) = 450$ in taxes. Thus it really only costs him $1350 to make the $1800 investment.

Example 5.2: Judy starts an IRA at age 21 and deposits $1200 at the end of each year until her 65th year. If her IRA earns 8.5% compounded annually, what is the value of her IRA after her last deposit?

> Solution: This is a typical regular payment problem, with $\mathbf{R}$ = 1200, $\mathbf{r}$ = 0.085, and $\mathbf{n}$ = 44.
>
> $$A = \frac{1200[\ (1+0.085)^{44} - 1]}{0.085} = \$497{,}176.48$$

Certificates of Deposit and Money Market Accounts

A Certificate of Deposit (CD) is issued to you when you deposit money in a special interest bearing account for a specified period of time. The time period can vary from 6 months up to five years, with higher interest rates on the longer CD's. Because the bank knows they will have your money for the whole period, they usually offer higher interest rates than on their savings accounts. There is a substantial interest penalty for withdrawing your money early, so you should be fairly sure you won't need it for the duration of the CD. An alternative to a CD is buying shares in a money market account. From the investor's point of view there is little difference between CDs and money markets; the interest rates and minimum deposits are usually very similar. The difference lies in what the bank does with the money. CDs can be thought of as individuals loaning money to the bank. Money markets can be thought of as individuals pooling money together through the bank to buy more expensive certificates such as jumbo CDs and Treasury bills. CDs and money markets are also subject to the $100,000 FDIC insurance limit.

Example 5.3: John has saved up $2500 to make a down payment on a car, but he won't be buying it until this time next year. His bank has CD's for 12 months which pay 5.35% compounded daily (365 per year). If he puts the money in a CD for a year, how much will he have for the down payment?

> Solution: John is making a lump sum deposit in an account paying 5.35% interest compounded daily for one year, where $\mathbf{P}$ = 2500, $\mathbf{r}$ = 0.0535/365 = 0.00001466, and $\mathbf{n}$ = 365 x 1 = 365.
>
> $$A = 2500(1 + 0.0001466)^{365} = 2500(1.0549528) = \$2637.38$$

Savings Bonds

The federal government offers series EE and HH savings bonds. EE bonds come in denominations (face value) of $100, $200, $500, $1000, $5000, and $10,000, and may be purchased at half the face value from financial institutions and many employers. In the past, bonds have not been competitive with other investments, but in 1982 the government changed the interest accrual method, and now they are very safe investments with a respectable yield. The interest on EE bonds is tied to the average yield on 5 year Treasury bills and is tax deferred. There is a set term to maturity, i.e., when the bond is actually worth face value, but interest only accrues for a total of thirty years. Mature EE bonds can be traded for HH bonds, on which interest is 7.5% paid semi-annually for up to 20 years. Bonds can be cashed in at any time without penalty. However, EE bonds cashed in too early may not yet be worth face value.

Effective Interest Rate (Yield)

When John called his bank about the interest rates on a Certificate of Deposit, he listened to a recording which told him that a 12 month CD of between $500 and $10,000 was currently available at a rate of 5.35% compounded daily. But immediately after that the voice said, "with a yield of 5.50%". This puzzled John so he called the bank and made an interesting discovery. The bank representative told John that because of compounding more frequently than once a year, the actual amount of interest his money earned was a little more than that 5.35%, in fact, it represented an actual or "effective" rate of 5.50%. She offered the following example:

If $100 is deposited at 8% interest compounded annually, the result after one year would be:

$$A = 100(1 + 0.08)^1 = \$108.00,$$

which represents $108 - $100 = $8.00 interest, an increase of **8%**. But if $100 is deposited at 8% compounded quarterly, the result after one year would be:

$$A = 100(1 + 0.02)^4 = \$108.24,$$

which represents $108.24 - $100 = $8.24 interest, an increase of **8.24%**. The actual increase of 8.24% is slightly higher than the stated increase of 8%. To keep these two numbers straight, 8% is called the **nominal** or stated rate of interest, while 8.24% is called the **effective rate** or **yield**.

So John says, "OK, let me try it. If the stated rate of the CD is 5.35%, then I could compare what happens with $100 again to find the effective rate, or yield." If $100 is deposited at 5.35% interest compounded annually, the result after one year will be:

$$A = 100(1 + 0.0535)^1 = \$105.35,$$

which represents $5.35 interest, an increase of 5.35% (as expected). But if $100 is deposited at 5.35% interest compounded daily, the result after one year will be:

$$A = 100(1 + 0.00001466)^{365} = \$105.50,$$

which represents $5.50 interest, an increase of 5.50% (actual yield). And that's exactly what the voice on the tape said, " ...a rate of 5.35% and a yield of 5.50%". The effective rate is useful in comparing the actual interest yield of various rates at various compoundings. Banks are required to tell you the actual yield on investments, so that comparison shopping is easier.

Example 5.4: John read an advertisement in the paper about a 12-month Certificate of Deposit which was offered at 5.4% compounded semiannually. What is the actual yield of that deposit?

Solution: $r = 0.054/2 = 0.027$, $n = 2$ gives

$$\text{yield} = 100(1 + .027)^2 - 100$$
$$= 5.47\%$$

Stocks, Bonds and Commodities -- Investing with Risk

The investments discussed above all have one basic feature in common: They have minimal risk (provided the government stays solvent!). They also can be characterized by their relatively low yields. Higher yields are possible if the investor is willing to shoulder some risk. For younger investors some risk is worth the potential high yields; investors nearing retirement should start limiting their risk by moving high risk investments into low risk investments. The most common forms of risky investments are stocks, bonds and commodities, also called **securities**. These investments are not insured. Moreover, not only might they not earn any interest, but initial investments can also be lost if business is bad.

Stocks

A stock is a share in the ownership of a corporation. The investor is called a shareholder. Stocks earn a portion of the company's profits – called a *dividend* – usually annually or quarterly. Many corporations offer a dividend reinvestment plan in which interest is automatically used to purchase more shares in the company. Most corporations offer two kinds of stocks: **Common stock** share holders receive a dividend based on profits and can vote on certain issues; **Preferred stock** shareholders have no voting privileges, but get a higher fixed annual dividend and higher priority for payment than common stocks. If the company's earnings are poor, stocks can end up being worth less than the original investment.

One special, less risky type of stock is a **utility stock**. A utility is a company that has been given the status of a state-regulated monopoly, e.g., electricity, gas, mass transit, and telephone companies. These stocks tend to be lower yielding but extremely stable.

Bonds

A bond is an IOU issued by a corporation (or government agency) promising to repay the investor with a given interest rate and term. In other words, the investor is loaning money to the bond issuer. Bonds have a higher priority for payment to investors than stocks. Two special forms of bonds are worth noting in particular: **Junk bonds** have an unusually high risk with an unusually high yield. **Municipal bonds** (sold by federal, state and local governments) are usually less risky and also tax exempt.

Commodities

A commodity is a product bought and sold on an exchange, e.g., gold, silver, wheat, corn, and pork bellies. Commodities can be invested in like stocks, but are considered higher risk. The most common way to invest in a commodity is to purchase futures: Buying or selling a commodity at a set time in the future for a fixed price, on the hope that the price will turn out to be lower (if buying) or higher (if selling) at that time. Obviously, unless you know something about the product's future availability, this is extremely speculative, and hence extremely risky. It is possible to lose more than the initial amount invested here, meaning that not only won't you have your principal back, but you could end up owing someone more money! How's that for high risk?

Mutual Funds -- A Compromise?

A mutual fund is an investment company that pools the funds of many individuals and invests the funds in a diverse set of low and high-yield stocks, bonds, commodities, etc., called a portfolio. Mutual funds are considered an excellent alternative to individual investments, since diversification often gives stability than individual securities. The SEC regulates all mutual funds, requiring them to publish portfolios and earnings track records. Investors can use this information to choose between funds.

There are many mutual fund companies to choose from. Often the only obvious difference is in how fees are charged to shareholders. The three main types of funds are No Load, Front-end Load and Back-End Load. In a no load fund there is no commission on purchasing or selling shares; investors pay an annual service charge. In a front-end load fund, investors pay a fee up front for buying shares (often 6% of the price of the shares) and usually no service charge. In a back-end (or rear-end) load, investors pay a fee at the end for selling shares.

Watching your Stocks

Stock information can be obtained from many sources. Stock listings, available in many newspapers, are an excellent source of information but take a little training to decipher. Here is a typical listing:

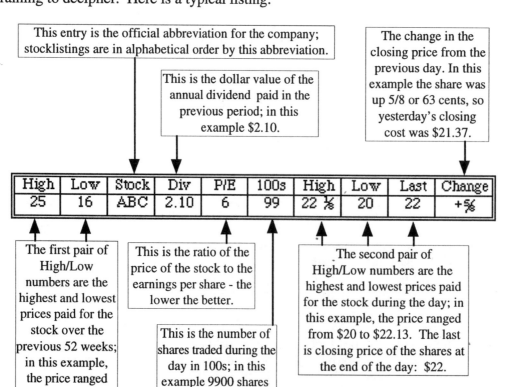

This entry is the official abbreviation for the company; stocklistings are in alphabetical order by this abbreviation.

This is the dollar value of the annual dividend paid in the previous period; in this example $2.10.

The change in the closing price from the previous day. In this example the share was up 5/8 or 63 cents, so yesterday's closing cost was $21.37.

High	Low	Stock	Div	P/E	100s	High	Low	Last	Change
25	16	ABC	2.10	6	99	22 ⅛	20	22	+⅝

The first pair of High/Low numbers are the highest and lowest prices paid for the stock over the previous 52 weeks; in this example, the price ranged from $16 to $25.

This is the ratio of the price of the stock to the earnings per share - the lower the better.

This is the number of shares traded during the day in 100s; in this example 9900 shares were sold.

The second pair of High/Low numbers are the highest and lowest prices paid for the stock during the day; in this example, the price ranged from $20 to $22.13. The last is closing price of the shares at the end of the day: $22.

These stock statistics can be used to help you make decisions about which stocks to invest in. Particularly useful are the price to earnings ratio (P/E) and the yield (dividend/current price or Div/Last). A low P/E ratio says that the company has high earnings relative to the price of the stock, so the stock is undervalued. This has two advantages: The price will likely rise so you will have bought in at a lower value, and if the market has a slump the price shouldn't sink as much as others. Conversely, a high P/E ratio indicates that earnings are expected to rise, i.e., that the company has shown promise and therefore the price of the stock has been driven up prematurely, and so is over valued. This can mean that the stock is over valued. How low is low? Look for P/E ratios below 7. In this example, the PE is 6, which would be considered low.

The ratio of Div/Last, which is 2.10/22 = 9.55% in this example, gives an indication of the yield as a percentage of your investment. Be careful when comparing stocks, the time period may be quarterly, semi-annually or annually. Check to be sure you are comparing similar periods.

Example 5.5: Here is a sample stock listing. Answer the questions below.

High	Low	Stock	Div	P/E	100s	High	Low	Last	Change
18	9	WW	0.96	4	12	17	15	15	-1

    a) Where did this stock close today? Yesterday?
    b) Is this stock under valued or over valued?
    c) How many stocks were sold during the business day?
    d) What is the yield of this stock currently?
    e) Has this stock been volatile in the last year? During the business day?

Solution:
    a)   Today the stock closed at 15 and yesterday at 15 - 1 = 14.
    b)   The P/E ratio is low at 4 so the stock is probably undervalued.
    c)   The 100s = 12, so 1200 shares were sold.
    d)   The current yield is Div/Last = 0.96/15 = 6.4%.
    e)   The share price has varied from $9 to $18 over the last year - which is a 100% increase; that seems volatile. During the day, the price varied between 15 and 17, which also seems to be a lot.

Before ending our discussion about investing in stocks, it is important to say a few words about the Stock Market. The market is said to reflect the state of the economy through past business success and future expectations. Many people rely on the Dow Jones Industrial Average to gauge how the stock market is doing. The Dow, as it is called for short, is a weighted sum of the stock prices of a mere 30 companies out of the thousands of stocks that are traded on various stock exchanges. These thirty companies are mostly older, more industrial companies like Proctor and Gamble, and so the impact of newer, high-tech companies is not reflected. There are other averages that incorporate many more companies, which perhaps makes them better indicators. Popular indices include Standard and Poor's Composite Index (S&P 500) with 500 companies, the National Association of Securities Dealers Composite (NASDAQ) with 3900 companies, and The Russell 3000 with 3000 companies.

YOU TRY IT 5.1
Answer these questions using the stock listing below.
    a) Where did this stock close today? Yesterday?
    b) Is this stock under valued or over valued?
    c) How many stocks were sold during the business day?
    d) What is the yield of this stock currently?
    e) Has this stock been volatile in the last year?
During the business day?

High	Low	Stock	Div	P/E	100s	High	Low	Last	Change
$20\frac{1}{4}$	$13\frac{1}{4}$	ARG	4.10	30.8	35	$15\frac{5}{16}$	15	$15\frac{3}{8}$	$-\frac{1}{8}$

Exercise Set 2.5

1) If George deposits $1000 in an IRA which pays 9% interest compounded monthly when he turns 32, how much will that $1000 be worth when he turns 65 years old?

2) George deposits $1000 in an IRA which pays 9% interest compounded monthly when he turns 32. The following year George deposits $2000 in the same account and then makes no further deposits. What will be the total value of his IRA when he turns 65?

3) Lucy is a Junior in High School. She has been saving for college and has $3000 at this point. Since she won't need the money for two years yet, she is considering putting the money in a CD that pays 6.2% interest compounded quarterly. If she does, how much will she have in two years?

4) Referring to Lucy in Exercise 3: Lucy's current savings account pays 4% interest compounded monthly. How much more will she have in two years if she buys the CD than if she left her $3000 in the savings account?

5) What is the effective rate or actual annual yield of a 24-month CD that pays 6.2% interest compounded quarterly?

6) You have $3000 to invest for two years. Which is better: A CD that pays 6.2% interest compounded quarterly or a CD at 5.9% interest compounded daily,?

7) Here is a sample stock listing. Answer the questions below.

High	Low	Stock	Div	P/E	100s	High	Low	Last	Change
32	20	LLL	0.10	7	16	$21\frac{1}{2}$	20	$20\frac{3}{4}$	$-\frac{3}{4}$

a) Where did this stock close today? Yesterday? What is the percent change in the price from yesterday to today?
b) Is this stock under valued or over valued?
c) How many stocks were sold during the business day?
d) What is the yield of this stock currently?
e) Has this stock been volatile in the last year? During the business day?

8) Here is a sample stock listing. Answer the questions below.

High	Low	Stock	Div	P/E	100s	High	Low	Last	Change
$73\frac{1}{2}$	$52\frac{3}{4}$	PPG	0.36	18.3	249	$75\frac{7}{8}$	$74\frac{7}{8}$	75	$+2\frac{13}{16}$

a) Where did this stock close today? Yesterday? What is the percent change in the price from yesterday to today?
b) Is this stock under valued or over valued?
c) How many stocks were sold during the business day?
d) What is the yield of this stock currently?
e) Has this stock been volatile in the last year? During the business day?

9) Compare the following two stocks. Be sure to mention yield and volatility. If you were planning on investing in only one, which would you choose. Note: ABC and XYZ both pay dividends quarterly.

High	Low	Stock	Div	P/E	100s	High	Low	Last	Change
$87\frac{1}{2}$	$80\frac{7}{8}$	ABC	0.71	16.3	732	$80\frac{7}{8}$	80	$80\frac{1}{2}$	$+\frac{5}{8}$
$83\frac{13}{16}$	$63\frac{3}{4}$	XYZ	0.57	20.5	422	$79\frac{1}{2}$	$78\frac{1}{2}$	$79\frac{5}{16}$	$+\frac{15}{16}$

10) Compare the following two stocks. Be sure to mention yield and volatility. If you were planning on investing in only one, which would you choose. Note: ABC pays dividends quarterly and XYZ pays dividends annually.

High	Low	Stock	Div	P/E	100s	High	Low	Last	Change
$25\frac{1}{2}$	$13\frac{1}{8}$	ABC	0.07	15.3	2	$22\frac{7}{8}$	$22\frac{7}{8}$	$22\frac{7}{8}$	0
$89\frac{5}{8}$	62	XYZ	0.25	30.6	452	$81\frac{1}{2}$	$80\frac{11}{16}$	$81\frac{3}{16}$	$+\frac{3}{16}$

CHAPTER SUMMARY

6

In this chapter we have discussed many ideas related to personal finance. The calculations involved in financial planning require little more than basic arithmetic and manipulation of exponents, so anyone with the access to the right formulas can do some of their own planning.

Key Ideas

Percents

- Percents are symbols representing the ratio ?? : 100.

- Percent increase (markup) can be calculated as new = old $\times (1 + \%)$.

- Percent decrease (markdown) can be calculated as new = old $\times (1 - \%)$.

Saving and Investing Money

- Lump sum investments are governed by the formula:

$$A = P(1+r)^n$$

- Ordinary annuities are governed by the formula:

$$A = \frac{R\left[(1+r)^n - 1\right]}{r}$$

- In general, the earlier you can start saving, the better because interest compounds.

- Many investments are tax deferred or tax exempt, making them attractive tax shelters.

- IRAs can be either lump sum investments or annuities, so choose a formula according to the situation.

- CDs are lump sum investments in which early withdrawal is penalized.

- Deposits made with banks and credit unions are insured under the FDIC.

- Securities – stocks, bonds and commodities – are not insured, and initial investments may be lost in addition to any earnings. They are regulated by the SEC.

Borrowing Money

- Credit cards accrue interest in different ways, the three main being ADB excluding new purchases, ADB including new purchases and 2 Cycle ADB.

- Loans are governed by the formula

$$R = \frac{rP}{\left[1 - (1+r)^{-n}\right]}$$

- Amortization tables have the form

Pmt #	Pmt Amount	Interest	Principal	Balance
	loan formula	r * prev. bal	pmt - interest	prev. bal - princ

- Debt to Income Ratio = monthly debt / monthly income. Typical ceilings are 40% total, 28% mortgage.

- Leasing a car can be an attractive alternative to buying. Leasing is essentially paying for the part of the life of the car used.

- Mortgages have more intricate closing costs, including bank, appraiser and attorney fees. In addition, lenders may require points, which are 1% of the principal and are usually tax deductible, as is all interest paid on a mortgage for your primary residence.

Summary Exercises

1) When politicians talk about the federal budget deficit, the words used can seem strange and the numbers involved are huge.

   a) The budget for 1992 was $1,381,895,000,000 and there was a deficit of $290,204,000,000 for that year. What percent of the budget was the deficit? If your family's budget was $30,000 per year and you had the same percentage deficit, how much money would you have over spent?

   b) In 1993, the budget was $1,408,122,000,000 and the deficit that year was $254,948,000,000. What percent of the budget was the deficit?

   c) What percent did the annual deficit drop from 1992 to 1993? What happened to the total deficit? (total deficit = $4,064,600,000,000 in 1992)

2) A typical ten speed bicycle has 2 sprockets (connected to the pedals) and 5 cogs (connected to the rear wheel). The chain can be wrapped around any cog - sprocket combination resulting in 10 different "speeds." As the cyclist pedals, he directly turns the sprocket, which drives the chain that turns the cog on the rear wheel. The table below gives the number of teeth on each cog and sprocket.

size	small sprocket	large sprocket	small cog	second cog	third cog	fourth cog	large cog
number of teeth	39	51	15	17	20	24	28

   The ratio of sprocket teeth to cog teeth gives the number of rotations of the rear wheel for one turn of the pedal. Calculate the ratios of sprocket teeth to cog teeth, and hence the number of revolutions of the rear wheel for one turn of the pedals, for each combination of sprocket and cog. Which combination is the fastest, i.e., pushes the cycle farthest? Slowest?

3) Joe decides to deposit his end-of-year bonus of $1000 into a retirement account which pays 6.44% compounded annually. Bob decides to spend his bonus on Christmas vacation each year, but he also deposits $1000 each year split into payments into a retirement fund. Bob makes equal deposits at the end of each month (what monthly payment will total $1000 in a year?). His account only pays 6.1% compounded monthly. Each man claims that in 10 years time he will have more in retirement. Who is right and why?

4) Pauline is 25 years old. She just opened a retirement account, and deposits

$100 at the end of each month into the account which pays 5.8% interest compounded monthly. Pauline's father, Paul is 45. He, too, just opened a retirement account in which he deposits $400 at the end of each month. Paul's interest rate is the same as his daughter's. Paul remarked to his daughter that he will have a much bigger retirement amount when he retires at age 65 (20 years from now) than she will when she retires at 65 (40 years from now). Is this true?

5) Ronnie wants to have a million dollars when he retires. If he is now 16 and plans to make deposits of $100 at the end of each year into an account paying 4.3% compounded annually, how old will he be when he retires?

6) Elizabeth deposits $200 at the end of each month into an account paying 6.5% interest, compounded monthly, for 3 years. She then leaves that money alone, with no further deposits, for 5 more years. What will be the final amount on deposit after the entire 8 years?

7) John deposited $12,800 in an account paying 7.48% interest compounded quarterly for 5 years. After the fifth year he found a better deal, so he emptied that account and deposited the total amount in an account paying 9% interest compounded semiannually. If he then left it in the second account for the next 4 years, what would be the total value of his account at the end of that time?

8) David deposits $500 at the end of each quarter into an account paying 6.5% interest compounded quarterly for 7 years. He then changes his deposit to $725 each quarter for 5 more years at the same rate. What will the amount on deposit be after the entire 12 years?

9) Angie has just placed $1000 in a savings account paying 7.25% interest compounded semiannually. She plans to leave it in the account for two years. That same day she reads about a savings account across town which pays 7.75% interest compounded daily (365 times a year). She figures it will take her about two hours to go to the first bank, get the money, go to the second bank, open the account, and go back to work. Since she has to take time off work to do it, it will cost her $6.50 per hour away from work, plus $1.50 in bus fare. Is it worth it for her to switch accounts? Why or why not?

10) Build the three amortization tables for the loan options given below.

1. A fixed rate 30 year mortgage at 7.5% payable monthly
2. A fixed rate 15 year mortgage at 7.25% payable monthly
3. A 5 year ARM at 6.4% payable monthly in which the interest may be raised or lowered no more than 2% every five years -- with a cap of 12%.

a) Consider the consequences of putting 15% down instead of 20% and paying PMI insurance calculated as 4% of the monthly payment.

b) Add to your work in (a) the escrow account, assuming taxes and insurance work out to be roughly $2000/year.

c) Add to your work in (b) the option of making extra payment towards the principal each month. What happens with $100? $200?

3 STATISTICS

Every day we are bombarded with data by the media. Variations in the experiments or observations that produced the data introduce error or **uncertainty** into the conclusions drawn from the data. **Statistics** is the science that deals with collecting, describing, and drawing inferences from data, and measuring the resulting uncertainty. In this chapter we will learn how to apply elementary statistical techniques to data and how to recognize misrepresentations of data.

After your work on statistics you will be able to organize data sets, display them graphically, and give summary information about them. You will also recognize good and bad graphs of data, and well done and flawed surveys. You will be able to graph data with two variables by hand and on the computer, find equations of lines approximating the points, and use those lines to make predictions.

History of Statistics

Drawing conclusions from data has been going on since prehistory, but before the eighteenth century there were only isolated cases that systematically dealt with uncertainty. For example, in 1100 AD. the London Mint used what we now call "random sampling" to maintain integrity of the coinage produced. The need for well defined techniques arose from difficulties in handling variability in astronomy, and were spurred on by interest in games of chance. Correspondence in the mid seventeenth century between the French mathematicians Fermat and Pascal debated the mathematics of gaming strategy. These letters contain the earliest discussions of probability, the theoretical foundation for present day statistical techniques. During the eighteenth century, many talented people were hard at work on formalizing statistics and probability.

By the mid nineteenth century, statistical approaches to collecting and analyzing data were well understood, and were widely used in the natural sciences. In the early twentieth century, the use of statistics spread to psychology where it became the main tool for researchers. More recently, statistical approaches have been applied to genetics, agriculture, manufacturing, economics and sports. Statistics has even been used in English literature; works attributed to Shakespeare were statistically analyzed, and the results were used to settle disputes about authorship. In fact, valid statistical analysis of data is pervasive in all disciplines from the humanities to business to the natural sciences. Statistics is everywhere!

COLLECTING DATA

1

Before drawing conclusions from data, we have to collect some. This step in the process is the most important and the most easily mishandled. Even the most sophisticated data analysis tools will draw nonsensical conclusions if given data that were not carefully and thoughtfully collected. Let's define some of the terms associated with data collection.

Population: The entire group of individuals or objects we are studying.

Characteristic: The specific question we want to answer or quantity we want to estimate regarding the population.

YOU TRY IT 1.1

Consider the following fictitious headlines and define the population and characteristic of interest for each:
a) Average SAT Score for ASU Incoming Freshmen up in 1993
b) Married Students at ASU Have Higher GPA than Singles
c) North Carolinians Prefer Pepsi Over Coke
d) Lifetime of Goodyear Tires is More Than 30,000 Miles

To gain information about the characteristic of interest, we can study the entire population or just a small portion of that population.

Taking a Census: Collecting data from the entire population.

Collecting a Sample: Collecting data from a subset of the population.

There are pros and cons for both of these collection methods. Studying the entire population will eliminate uncertainty in the results, but may be too time-consuming, costly, or impractical. For example, suppose that we want to determine the average length of all fish in Watauga Lake. All fish in the lake would comprise the population, and the average length is the characteristic of interest. We could actually measure the length of any fish in the lake, but it is not practical to take a census, i.e., to remove and measure all fish. Instead, we will only remove and measure a sample of 50 fish from Watauga Lake. This sample of 50 measurements can be used to draw inferences about the entire population of fish, provided the sample is chosen correctly.

There are several agencies that gather data using a census. For example, the Federal government gathers data about the entire population of the United States every ten years. However, this is not the most common method of data collection. We will concern ourselves with the more widely used technique of sampling for the rest of the section.

Sampling

Proper choice of sample is crucial if the sample is to be truly representative of the entire population. **Sample bias** is error caused by improperly choosing a sample. Sampling individuals based on friendliness, appearance, or convenience will cause some parts of the population to be under-represented and others to be over-represented. Statisticians make great efforts to eliminate bias in the chosen sample to be measured. Let's look at several common ways bias might be introduced into sampling.

Sampling the members of a population that are easiest to reach is called a convenience sample. **Convenience samples** usually produce data that are not indicative of the responses or feelings of the entire population. When gathering data on the percentage of ASU students who own a personal computer, we might question every student who enters Walker Hall. We might mistakenly assume that an elderly person or a middle-aged man who had on work clothes is not a student. Furthermore, only mathematics, computer science, statistics, and communications courses are taught in Walker Hall; therefore, many upper level ASU students with other majors would come to Walker Hall infrequently. Our convenience sample data would not accurately reflect the percentage of ASU students who own a personal computer.

Voluntary response is another typical source of bias in sampling. Suppose a nuclear power generating plant is to be built on the outskirts of our county. A radio newscaster asks his audience to express their views on the subject by means of a call-in poll. Typically people with strong feelings, usually negative feelings, will call in. In this case the people sampled select themselves. Voluntary response sampling generally results in a poorly chosen sample.

Personal choice may also cause sampling to give unreliable results. Some polling companies administer questionnaires that take an hour to complete. Pollsters go to houses in randomly chosen blocks to get people to fill them out. If a pollster is more likely to approach an older person rocking on the front porch than a mother taking care of several children, she would be allowing personal choice to bias the sample.

Example 1.1: One of the most famous cases of sampling bias occurred during the 1936 presidential campaign. Franklin D. Roosevelt, incumbent President, was running against the Kansas Republican Governor Alfred Landon. The *Literary Digest*, a very respected magazine had successfully predicted presidential election results since 1916. It chose its sample from telephone directories, magazine subscription lists, and club membership rosters. Ten million ballots were sent out, and 2.4 million were returned. The *Digest* predicted that Landon would receive 57% of the vote to Roosevelt's 43%, but Roosevelt won by a landslide. What happened?

> Solution: There were two main sources of error in the *Digest* sampling technique: convenience sample bias and voluntary response bias. This sample clearly leaves out the poor and uneducated. This election occurred during the Great Depression, a time when telephones, magazines, and club memberships were quite a luxury. Voluntary response bias also played a role since less than 25% of those polled responded.

Random Sampling

One way to overcome bias in sampling is to use **random sampling**; Choose the sample so that any member of the population has the same chance of being in the sample as any other member. If we were sampling the lengths of 50 fish from Watauga Lake, then every fish from the tiniest minnow to the largest catfish must have an equal chance of being in the sample.

Suppose that we have 20 students. From those 20 students we want to randomly select 10 for a special project. We could put the names of all 20 students in a hat, mix them well, and then draw out 10 names. These 10 would be our random sample.

Sometimes the population size is too large to put names of all members of the population into a hat. Suppose our population contains 100 students instead of just 20. In this case we can assign a number to each member of the population and then use a **table of random digits** to choose our unbiased sample.

Let's draw a random sample from the population of 100 students to illustrate how to use that table. Number the 100 students using 2-digit combinations from 00 to 99. (We could start with 1, but then we'd need 3-digit combinations for 001 to 100.) Start in any row of the table on the next page and write down successive groups of 2 digits, ignoring any 2 digit group that does not give a number between 00 and 99, or any 2 digit group that has already been selected. Don't worry about spaces in the rows; they're there to help you keep track of where you are in the row. For example, starting on line 22 (any line will do), we get the numbers:

$$20, 63, 63, 35, 80, 86, 72, 00, 18, 91, 87.$$

We crossed out the repeat 63. Our sample of ten students out of 100 consists of those students whose numbers appear in this list.

TABLE OF RANDOM DIGITS

ROW	C1	C2	C3	C4	C5	C6
1	38454	96671	49592	95692	95796	40099
2	50939	38245	78127	82719	55164	33411
3	42956	65266	31388	27024	91326	24384
4	33609	61417	29124	19419	78225	25506
5	95173	67776	40305	76200	88522	86112
6	57511	92303	32148	92745	62331	91435
7	09203	79258	84880	48833	29825	66272
8	71952	91801	06207	26199	30409	90618
9	63447	56868	76374	20048	48536	51555
10	16232	10207	47366	86688	65728	89049
11	87901	51354	14340	51510	24488	71540
12	11583	80379	27725	69008	11407	15834
13	66063	21571	50877	21030	54993	59155
14	91913	83463	64818	21955	94995	86218
15	97211	29863	10563	73921	87152	62095
16	24686	09029	91294	70610	47528	10200
17	34661	51675	05791	31385	90427	14824
18	52320	38935	77820	18891	93546	24114
19	82158	34619	44895	56988	03751	43639
20	37532	95600	57256	02401	58720	14846
21	10251	88138	31625	24024	70382	03541
22	20636	33580	86720	01891	87763	31999
23	06671	16503	71673	11043	04652	83694
24	41623	51028	35181	87251	28166	86191
25	17207	94295	09034	31713	02349	82645
26	19532	81650	96885	53113	73561	47792
27	33886	28193	38971	86018	04560	40981
28	67976	70052	59532	47884	84547	78464
29	37152	20876	69057	37654	49511	37226
30	58450	68497	24966	98627	27881	89937
31	72501	20120	56197	00415	74898	08175
32	90046	99091	19264	77860	87785	94224
33	17504	95388	92269	63986	05089	30848
34	65768	54861	52607	60503	68348	22919
35	04369	37765	11202	70197	94422	55843
36	52075	18012	23623	23484	93672	51518
37	67728	02446	08736	17851	28367	55850
38	74967	76976	40634	31315	17941	36246
39	81110	07776	92619	39581	31388	39491
40	25218	42085	16612	95631	74030	16245

YOU TRY IT 1.2

Criticize the following sampling techniques and decide what is good and bad about each one:

a) Jill stands outside the cafeteria's main entrance and asks students whether or not they are satisfied with current parking regulations.

b) Dave stands out in the lobby between classes and asks students and faculty whether or not smoking should be allowed in public buildings.

c) A survey asking whether or not a person believes cigarettes should be more heavily taxed is placed in every post office box in the Boone Post Office. People are asked to mail back their responses.

d) Mandy goes to every household within a 10-block radius of city hall and asks citizens whether price supports should be continued for farmers.

Example 1.2: Suppose we want to randomly choose 10 students from the 872 in the senior class. Describe how to do this.

Solution: We will assign a number to each of the people in this population. Three digits are needed to label the 872 seniors. We number from 001 to 872. Using the Table of Random Numbers starting on line 16, we get the numbers:

246, 860, ~~902~~, ~~991~~, 294, 706, 104, 752, 810, 200, 346, 615.

We crossed out 902 and 991 because they were too big, and we just continued on the next line when we reached the end of line 16.

YOU TRY IT 1.3

Use the Table of Random Digits to choose a random sample of 8 cars out of a group of fifty cars that have been numbered from 1 to 50.

Variability in Sampling

Suppose that when we randomly chose 10 people from the senior class in Example 1.2, we chose 4 males and 6 females. When we do this a second time, suppose we end up with 3 males and 7 females. A third random choice yields 5 males and 5 females. Each of these random samplings gives a different percentage of females. Although random sampling eliminates bias, it certainly does not eliminate **variability** - the difference in results from one random sample to another.

You may wonder why we should use random samples if there is variation from one sample to another. The answer lies in the fact that there is a specific chance mechanism used in random sampling, and therefore the results are not haphazard over the long run. Repeatedly choosing random samples of 10 people from the senior class can be trusted to give results that on average are close to the senior class's actual male to female ratio, and the individual samples will vary from this ratio in a predictable fashion.

Example 1.3: Given the 15 random samples below of 10 people each from the 872 seniors, verify that the results are close to the 49% male to 51% female ratio on average.

Sample#	1	2	3	4	5	6	7	8	9	10	11	12	13	14	15
Females	5	6	6	5	5	3	5	6	4	5	5	7	6	5	4

Solution: Let's arrange these data in a graph called a histogram. (We will discuss histograms more thoroughly in the next section.) The graph merely shows how many times 3 females were selected, how many times 4 females were selected, and so on.

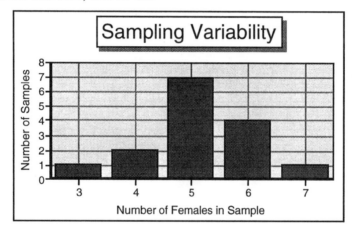

We can see from the heights of the individual bars that there were only 3 females in one of the samples, 4 females in two of the samples, 5 females in seven of the samples, 6 females in four of the samples, and 7 females in only one of the samples.

The graph above has the shape we expect from a set of random samples, high in the middle near the true population characteristic, sloping down almost evenly in either direction. Statisticians call this the **normal distribution**. In our example, if we were to take repeated random samples and observe what happens, we would find that most samples would contain about 5 females. In this way random sampling comes very close to predicting a characteristic of a given population. It must be said, however, that a sample size of 50 will come closer to the truth than a sample of size 10.

Statisticians often use random sampling to predict from a sample the characteristic of interest for the entire population. A good rule of thumb for simple statistical analysis can be stated as:

To estimate a characteristic of a population, take a reasonably sized random sample and use the sample characteristic as the estimate.

Surveys

Once we choose a sample, we must gather the data needed to gain insight into the characteristic of interest. If the characteristic is something that can be measured objectively, like length in inches of fish in Watauga Lake or amount of bacteria present in grams, it only remains to carefully and consistently measure each member of the sample. Often the characteristic is not an objective quantity, but rather a subjective feeling or opinion. To elicit this information from the sample, a set of questions in the form of a **survey** can be employed. Care must be taken in constructing questions in order to avoid biasing the responses of the members of the sample. Let's look at an example.

Example 1.4: During the health care debate of 1994, many congressmen resorted to surveying their constituents before voting on issues related to health care. Here are the first three yes/no questions asked on one such survey. What do you think this congressman really wanted to hear from his constituents?

1. Are you generally satisfied with the quality of the health care you receive?

2. Are you satisfied with the cost of your health care?

3. Do you believe that because 37 million Americans do not have health insurance, and because health care costs have been rising considerably faster than inflation in recent years, that we need major health care reform legislation?

> Solution: The first two questions on this survey are okay. The third question is highly suspect. The opening clause on people who don't have health insurance would probably lead most people to answer "yes" to this question.
>
> Most likely this is the answer the congressman wants his constituents to give. Leading questions like this one should be rephrased if he really wants to get honest opinions. Question 3 as written is more appropriate as "propaganda" than as a question on a survey.

The example above should give you an idea of how *not* to ask questions and how to spot when someone is trying to sway your answer to a question. Beware! Many lobbyists use this type of "survey" to influence public opinion.

Survey Questions
1. Watch for biased, leading questions. Using seemingly awkward wording like "Are you satisfied or dissatisfied with..." is a better way.
2. Put the most important questions first. Early questions might influence responses to later ones.
3. Keep the response list small and specific. Open ended questions are hard to handle; specific choices such as yes/no, always/sometimes/never, once a day/week/month/year, etc., are easier to analyze.
4. Test the wording for possible misinterpretations. Be as precise as possible.
5. Ask the questions or deliver the survey in exactly the same way to each member of the sample. Opening phrases, whether oral or in writing, should be standardized.

YOU TRY IT 1.4

The city council has been asked to change the zoning on a property to allow for a new Bigmart store. The council decided to conduct a poll to determine whether residents wanted a major discount store in town. Council members submitted the questions given below.

1) How often do you leave town to shop at a Bigmart or other discount store?
    often   occasionally   rarely   never

2) Would you shop at a Bigmart instead of other existing stores in town?
    often   occasionally   rarely   never

3) Bigmart has been accused of predatory pricing (forcing smaller businesses out by temporarily undercutting their prices). Would you be concerned about how Bigmart would impact local businesses?
        extremely   somewhat   no

Critique the questions, suggesting changes to increase the likelihood of an unbiased response.

Experimentation

Suppose that the question we want to answer deals with how a population will react to a certain stimulus, for example how corn plants will react to a fertilizer not yet available to the general public. We cannot reuse existing data, so we must design an **experiment**. This is different from sampling, because we will impose a treatment on members of the population in an effort to verify a cause and effect relationship. Such relationships can be algebraically explored using **regression**, as we'll see in Section 4. Experiments must be carefully designed to ensure that the conclusions drawn are valid. Here's an example.

Example 1.5: The FDA wants to test the effectiveness of a new cholesterol lowering drug. How should it proceed?

> Solution: In human testing, especially with subjectively measured changes, it is important to avoid the **placebo effect**; people given no treatment expect nothing to happen. To counteract this, all participants are given something. Some are given the drug and others the inactive placebo, none knowing which they received. Even those administering the medication should not know whether the drug or placebo is given. This style of experiment is called **double-blind**.
>
> Thus to test a new drug, the FDA might perform a double-blind experiment: Divide participants randomly into two groups, administer the placebo to one and the actual drug to the other, and record all results together after the appropriate time interval has passed.

The placebo effect is one example of **confounding**. Variables are said to be confounded when their effects cannot be clearly separated. Let's consider a corn fertilizer experiment. We give farmers in Iowa the new fertilizer and farmers in Texas an existing fertilizer. The Iowa farmers notice a bigger yield than their comrades in Texas. Can we conclude that newer is better? What about quality of soil, method of dispersal, length of season? These other possible factors must be eliminated before we can draw a conclusion.

Exercise Set 3.1

1) There may be a number of reasons for taking a sample instead of a census of an entire population. The Stateline Fireworks Company needs to know whether the fire crackers it produces will explode. It tests one out of every 500 to find out. What do you think is the main reason for using a sample instead of a census in this situation?

2) There may be a number of reasons for taking a sample instead of a census of an entire population. The Nielsen television rating service determines the popularity of national programs by a sample of 1200 homes out of all those American homes with a TV. What do you think is the main reason for using a sample instead of a census in this situation?

3) What could cause the results in each of the following situations involving sampling to be inaccurate?

a) The Madame Whiffit Perfume Company wants to know the average age of the ladies who buy its products. Madame Whiffit sends representatives to various stores to ask customers their ages.

b) In order to predict which candidate will win a presidential election, a national magazine sends ballots to all of its readers.

4) What could cause the results in each of the following situations involving sampling to be inaccurate?

a) Dr. Fetherbrane, a psychiatrist, wants to find out what percentage of people dream in color. He asks 100 of his patients whether they do or not.

b) A high school math teacher wants to find out if math anxiety is gender based. She asks all of her honors calculus classes whether they experience math anxiety when taking tests.

5)   In each of the following exercises briefly identify the population, the characteristic of interest, and the sample. If the situation is not described in enough detail to completely identify the population, complete the description of the population in a reasonable way. Furthermore, each sampling situation described in these exercises contains a serious source of bias. In each case, discuss the reason you suspect that bias will occur and the direction of the bias. (In what way will the sample conclusions differ from the truth about the population?)

a) The Chicago Police Department wants to know how black residents of Chicago feel about police service. A questionnaire with several questions about the police is prepared. A sample of 300 mailing addresses in predominantly black neighborhoods is chosen, and a police officer is sent to each address to administer the questionnaire to an adult living there.

b) A flour company wants to know what fraction of Minneapolis households bake some or all of their own bread. A sample of 500 residential addresses is taken, and interviewers are sent to these addresses. The interviewers are employed during regular working hours on weekdays and interview only during those hours.

6)   In each of the following exercises briefly identify the population, the characteristic of interest, and the sample. If the situation is not described in enough detail to completely identify the population, complete the description of the population in a reasonable way. Furthermore, each sampling situation described in these exercises contains a serious source of bias. In each case, discuss the reason you suspect that bias will occur and the direction of the bias. (In what way will the sample conclusions differ from the truth about the population?)

a) A national newspaper wanted Iowa's reaction to President Reagan's agricultural policy in early 1984. A reporter interviewed the first 50 persons willing to give their views, all in a single voting precinct. The headline on the resulting article read "Reagan Policies Disenchant Iowa," and the reporter wrote that Reagan would lose an election in Iowa if one were held then.

b) A congressman is interested in whether his constituents favor a proposed gun control bill. His staff reports that letters on the bill have been received from 361 constituents and that 283 of these oppose the bill.

7)   Are random number tables really random? To be random, each digit must be equally likely. Check the Random Number Table given in this section by choosing ten rows and counting the occurrences of each of the digits 0 through 9. Does each of the digits occur about the same number of times?

8)   Use the Table of Random Digits to select a random sample of 5 of the following 25 volunteers for an experiment on drug interaction. Be sure to say where you entered the table and how you used it.

Ackerman	Garcia	Petrucelli	Casella
Andrews	Healy	Reda	Frank
Baer	Hixson	Roberts	Filbert
Berger	Lee	Smith	Furman
Brockman	Lynch	Snell	Filmore
Carson	Moser	Turner	George
Wilson			

9)   On the next page is a population of 60 circles.

a) Compute the average diameter of the circles.

b) Close your eyes and drop a pencil onto the page, marking the circle you hit, until you have marked 4 circles. Is this a random sample? Compute the average diameter of your sample, and compare it with the actual average.

c) Label the circles 01, 02, ..., 60 and use the Table of Random Digits to choose a sample of size 4. Compute the average diameter of these 4 circles and compare with the average you found in b) and with the actual average.

d) Repeat (b) and (c) using a sample of 16 circles. How does this compare to the samples of 4 circles? Which do you think had less bias: the pencil drop method or the random number table method?

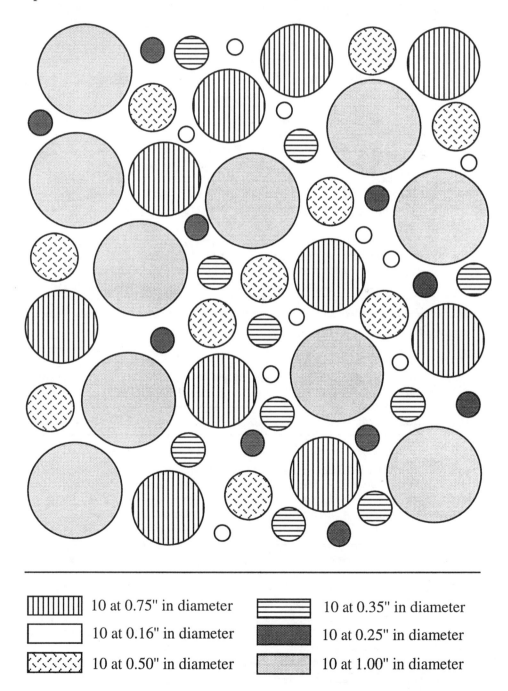

‖‖‖‖‖	10 at 0.75" in diameter	☰	10 at 0.35" in diameter
▯	10 at 0.16" in diameter	■	10 at 0.25" in diameter
⬚	10 at 0.50" in diameter	▢	10 at 1.00" in diameter

10)   Describe in detail how you would go about designing an experiment to determine the effect of the interaction of aspirin and caffeine when given to

patients who suffer chronic headaches. Be sure to consider compounding effects carefully.

11) To investigate sampling variability and the normal distribution, perform the following experiment. Put 30 m&m candy pieces in a cup. Draw out a random sample of 5 and record the percent yellow (number of yellow in the sample ÷ 5). Do this 15 times, returning the drawn pieces each time, and record your results. What do you think is the percentage of yellow m&m candies in the entire cup? Check how close you came.

12) The following is an excerpt from Mark Twain's Life on the Mississippi. The document contains 376 words. On the left are the numbers of words in each paragraph. Hyphenated words are counted as two and numbers as one.

a) Use the Table of Random Digits to draw a random sample of 5 words. Compute the average word length of your sample.

b) Draw a random sample of size 15. (Use a different part of the Table of Random Digits than used in a).) Compute the average from this sample.

58   *Since my own day on the Mississippi, cut-offs have been made at Hurricane Island, at Island 100, at Napoleon, Arkansas, at Walnut Bend, and at Council Bend. These shortened the river, in the aggregate, sixty-seven miles. In my own time a cut-off was made at American Bend, which shortened the river ten miles or more.*

66   *Therefore the Mississippi between Cairo and New Orleans was twelve hundred and fifteen miles long one hundred and seventy-six years ago. It was eleven hundred and eighty after the cut-off of 1722. It was one thousand and forty after the American Bend cut-off. It has lost sixty-seven miles since. Consequently, its length is only nine hundred and seventy-three miles at present.*

86   *Now, if I wanted to be one of those ponderous scientific people, and "let on" to prove what had occurred in the remote past by what had occurred in a given time in the recent past, or what will occur in the far future by what has occurred in late years, what an opportunity is here! Geology never had such a chance, nor such exact data to argue from! Nor "development of species," either! Glacial epochs are great things, but they are vague --- vague. Please observe:*

166   *In the space of one hundred and seventy-six years the Lower Mississippi has shortened itself two hundred and forty-two miles. That is an average of a trifle over one mile and a third per year. Therefore, any calm person, who is not blind or idiotic, can see that in the Old Oolitic Silurian Period, just a million years ago next November, the Lower Mississippi River was upward of one million three hundred thousand miles long, and stuck out over the Gulf of Mexico like a fishing-rod. And by the same token any person can see that seven hundred and forty-two years from now the Lower Mississippi will be only a mile and three-quarters long, and Cairo and New Orleans will have joined their streets together, and be plodding comfortably along under a single mayor and a mutual board of aldermen. There is something fascinating about science. One gets such wholesale returns of conjecture out of such a trifling investment of fact.*

13) Here are a few questions from a survey mailed to a random group of U.S. citizens by a major farming lobby. What do you think this lobby's position is on farm land legislation? Critique the survey and suggest any changes that would make the questions impartial. The possible answers are yes, no, and undecided for these questions.

1) It is estimated that the U.S. converts 2 million acres of farmland to development every single year. Is this acceptable?

2) If the current rate of farmland lost each year continues, the U.S. could become reliant on imported foods. Does this concern you?

3) Were you aware that the federal government spends $40 million dollars a day on farm subsidy programs?

4) We advocate the shift of one weekend's worth of farm subsidy payments to funding legislation that would help save farmland from development. Do you agree that this is a wise reallocation of funds?

14) Here are two questions from a survey mailed to a random group of town residents by a citizen's group. What do you think this group's position is on liquor by the drink? Critique the survey and suggest any changes that would make the questions impartial. The possible answers are yes, no, and undecided for these questions.

1) It is estimated that there is an alcohol related automobile accident on our streets every 5 minutes. Is this acceptable?

2) Communities in which liquor by the drink has been legalized have witnessed a 25% increase in alcohol related accidents. Do you favor legalizing liquor by the drink in our community?

15) The student government association wants to estimate the number of students who have been victims of violent crimes on campus in the past academic year. Design a survey for the SGA that would give it the information needed. How and to whom would you deliver your survey to get the best possible results?

16) The local public library wants to change its hours to be more convenient for county residents to check out books and use the library's other facilities. Design a survey for the library managers that would give them the information needed to set better hours. How and to whom would you deliver your survey to get the best possible results?

17) Does learning a foreign language improve English skills as well? High school students who elected to take French in high school were tested and their scores were higher than average. Explain how confounding might have played a role in this experiment.

18) Does the herb ginseng have restorative powers? A nurse served ginseng tea to nursing home residents several times a day, visiting with them while they drank. After a month, the residents were markedly more alert. Explain how confounding might have played a role in this experiment.

Presenting Data

2

Once a data set has been collected, we must decide how to present it in a readable, organized fashion. If the data set is very small, a complete list might be best; for data sets bigger than ten or fifteen numbers, a summary in the form of a picture will be easier to read and draw conclusions from quickly. Pictures like graphs and charts are common data presentation tools and have become ubiquitous in the media due to the increased use of computer graphics. As users of statistics we need to learn how to clearly present data; as consumers we need to learn how to interpret other people's graphical presentations of data. We'll examine some of the more common types of graphs and charts in this section, and look at some common mistakes to avoid.

Histograms

One way to display data is to use a histogram, which is based on a grouped frequency table. Each bar in a histogram represents one class or set of responses and is as long as the number of responses in that class. To see how this works we asked 25 math students to estimate how many hours a week they watch TV. Here are the data we collected:

Person:	1	2	3	4	5	6	7	8	9	10	11	12	13
Hours of TV:	10	3	6	8	20	35	4	20	3	15	5	20	25

Person:	14	15	16	17	18	19	20	21	22	23	24	25
Hours of TV:	3	28	0	14	2	1	0	5	5	30	4	0

Let's draw a histogram. There are three main steps involved.

1. **Divide the data into classes of equal width.**

   We will choose these classes:

0	$\leq$	# Hrs. TV	<	5
5	$\leq$	# Hrs. TV	<	10
10	$\leq$	# Hrs. TV	<	15
15	$\leq$	# Hrs. TV	<	20
20	$\leq$	# Hrs. TV	<	25
25	$\leq$	# Hrs. TV	<	30
30	$\leq$	# Hrs. TV	<	35
35	$\leq$	# Hrs. TV	<	40

   There is no correct number of classes in a histogram. You may need to experiment with different class sizes; your goal is to give a clear picture of the distribution. If you choose too few classes, you get a skyscraper effect where all observations are bunched together; choosing too many classes will give a flat histogram with too few observations in each class. A reasonable rule would be to use the number of classes necessary to show a peak in the distribution, while using no less than 5 and no more than 15.

2. **Make a frequency table.**

   We count and record the number of observations in each class. This is an optional step. For small data sets it is easier just to count as you draw the histogram.

Class					Frequency
0	≤	# Hrs. TV	<	5	10
5	≤	# Hrs. TV	<	10	5
10	≤	# Hrs. TV	<	15	2
15	≤	# Hrs. TV	<	20	1
20	≤	# Hrs. TV	<	25	3
25	≤	# Hrs. TV	<	30	2
30	≤	# Hrs. TV	<	35	1
35	≤	# Hrs. TV	<	40	1

3.   **Draw the histogram.**

The hours of TV per week is on the horizontal scale, and the frequencies are on the vertical scale. Each bar represents a class, and the bar height is the class frequency.

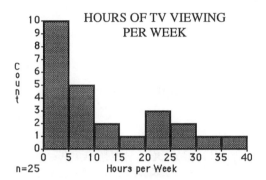

There are several things we can say about these data. It is apparent from the histogram that the "0 ≤ # Hrs. TV < 5" class contains more observations than any other. The majority of the data is in the "0 ≤ # Hrs. TV < 10" classes. The distribution of the data has a long right tail with no matching left tail.

YOU TRY IT 2.1

Construct a histogram from the following final exam scores for a sociology class.

74	83	70	55	76	75	59	92	56	50
57	70	52	74	77	83	55	78	13	74
77	55	51	70	57					

a) What should your classes be? Experiment with different classes.
b) Make a frequency table for the exam scores.
c) Draw the histogram and describe the distribution

Relative Frequency Histograms

A relative frequency histogram gives another way to display univariate data. On the vertical axis we use **relative frequencies** or percentages. Relative frequency histograms are very useful when comparing two data sets of differing sizes. Look back to the data on "Hours of TV Viewing per Week" on page120. Here is the frequency table again with the relative frequencies added. To get the percent we must divide the frequency for each class by the total number of observations in the data set, in this case 25.

		Class			Frequency	Relative Frequency
0	≤	# Hrs. TV	<	5	10	10/25 = 0.40
5	≤	# Hrs. TV	<	10	5	5/25 = 0.20
10	≤	# Hrs. TV	<	15	2	2/25 = 0.08
15	≤	# Hrs. TV	<	20	1	1/25 = 0.04
20	≤	# Hrs. TV	<	25	3	3/25 = 0.12
25	≤	# Hrs. TV	<	30	2	2/25 = 0.08
30	≤	# Hrs. TV	<	35	1	1/25 = 0.04
35	≤	# Hrs. TV	<	40	1	1/25 = 0.04

Now let's draw our relative frequency histogram.

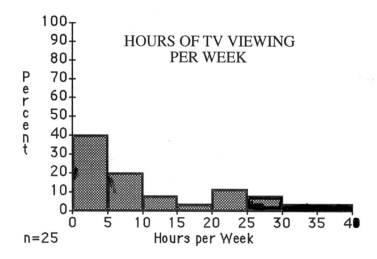

We see that 40% of the students watch less than 5 hours of TV per week, 20% of the students watch between 5 and 10 hours per week, and so on. Note that a student who watches exactly 5 hours of TV per week is in the "5 to 10" class and not the "0 to 5" class.

Suppose that we had data on the number of hours of TV watched per week from 100 music students. (The above data were math students.) In order to compare these data sets, we would have to deal with the fact that the sets are different sizes, 25 and 100. If 20 music students say they watch between 0 and 5 hours of TV a week, it is meaningless to say that twice as many music students as math students watch between 0 and 5 hours a week. We're comparing apples and oranges. To level the playing field, we must consider the relative frequencies: 40% of the math students say they watch between 0 and 5 hours a week as compared to 20% of the music majors, giving a totally different impression.

YOU TRY IT 2.2

The data below are first exam scores from a section of calculus. The data are separated into scores for people who had pre-calculus the previous semester and scores for those who went straight into calculus as freshmen. Construct two histograms and then two relative frequency histograms for these data, using the same classes for all graphs.

Had Pre-Calc:	62	71	72	75	77	81	88	92	95	97	
Didn't:	55	61	68	72	77	77	80	81	82	85	86	86
	89	89	90	91	94	98					

Discuss which pair of graphs would be better for a direct comparison of performance between those who had pre-calc and those who didn't.

0-60 ~1
0-70 -2
0-80 -4
0-90 ~8
0-100 -2

scores in intervals

Pie Charts

When the data we want to present show how something is subdivided, we can use a pie chart. The Federal government uses pie charts to illustrate how the annual budget is spent. Look at the last pages of any Form 1040 Tax Return instruction booklet for an example.

Example 2.1: A student is asked to illustrate how she spends a typical day, and she gives the following information.

Sleeping	8 hours	Classes	5 hours
Studying	5 hours	Watching TV	2 hours
Other	4 hours		

Build a pie chart to show the distribution of this student's time.

Solution: We need to calculate what percent of 24 hours she spends on each activity:

Sleeping: $8/24 \approx 0.333$ Classes: $5/24 \approx 0.208$
Studying: $5/24 \approx 0.208$ Watching TV: $2/24 \approx 0.083$
Other: $4/24 \approx 0.167$

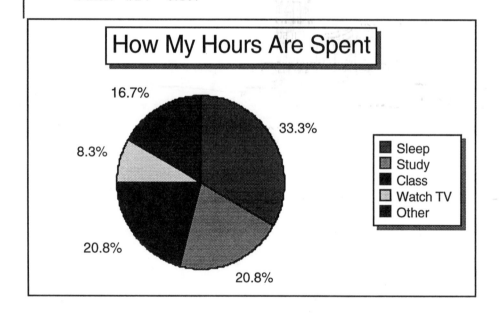

To draw a pie chart by hand, we would need divide the circle into angle of the appropriate size using a protractor. To determine the angle that corresponds to 33.3% in the example above, we calculate the angle that is 33.3% of 360°. Multiplying 360 by 0.333 gives 119.88 or about 120°. We calculate all the angles this way and then sketch the pie chart. In order to make pie charts easy to read, the wedges are usually in clockwise order starting at the top as listed in the legend.

Line Graphs

When the data we need to illustrate involve the change in a quantity over time, a line graph works well. Consider the U.S population data:

Year of Census	1940	1950	1960	1970	1980	1990
Population in Millions	131.67	151.33	179.32	203.21	226.50	249.63

If we want to show the trend over time we can plot the population for each year and connect the dots with straight lines. This gives us a nice visual of the trend in the population growth since 1940. We need to be sure to use equal spacing for the time intervals on the horizontal axis and the population scale on the vertical axis so that the slopes of the lines between points are not distorted.

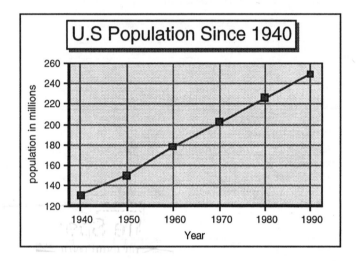

YOU TRY IT 2.3

Construct a line graph from the following population data for the United States. (The population is given in millions.)

Year	1790	1820	1850	1880	1910	1940	1970
Population	3.93	9.64	23.19	50.16	91.97	131.67	203.21

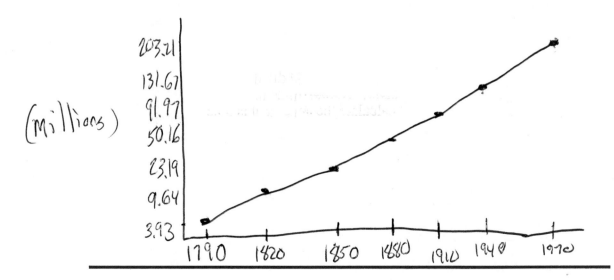

Tables and Cross Tabulations

When presenting the results of a survey, especially one with a small set of responses like yes/no or agree/disagree, a table recording the total number or percentages for each answer is usually the clearest way to present the data. Sometimes analysis of the responses shows a relationship between subgroups of people surveyed and their responses. A nice way to display these data to illustrate such differences is to build a table showing response totals for each subgroup and the entire group. For example, suppose we administer a survey to men and women to determine whether or not they agree with the President's stand on a particular issue involving gender. It might be natural to ask whether responses of men and women were different. Consider the table given below.

	women	men	total
yes	213	122	335
no	88	70	158
undecided	4	6	10
totals	305	198	503

From this **cross-tabulation**, we can see at a glance that more women than men participated in the survey, and that women gave more yes answers. To make it even clearer how women compared to men, and how the entire group responded, record the percentages calculated column-wise in the table:

	women	men	total
yes	70%	62%	67%
no	29%	35%	31%
undecided	1%	3%	2%
totals	100%	100%	100%

From this table, we get a clear picture of the pattern in the data. For example, 70% of women said yes, 3% of men were undecided, and 31% said no overall. It is interesting to note that the percentages for men and women don't differ that much from each other or from the overall average percent of people who answered yes, no or undecided; it would be fair to say that there were only minor differences between men and women on this issue.

We could have calculated the percentages across the rows instead (the last column would all be 100% in that case) if we were wondering what percent of the yes votes were women instead of what percent of women voted yes. It helps to decide which kinds of questions you're interested in before building the table.

Example 2.2: A survey was conducted in an effort to determine how residents felt about the building of a state correctional facility on the outskirts of the city. The city council also wanted to determine whether the property values of the residents made a difference in their opinions. The raw data are:

```
Total surveyed:   500 residents
                        in favor      opposed      no opinion
Property value ≥ $150,000:    12           72           16
Property value < $150,000:    60          240          100
```

Solution: Adding the numbers reveals that 100 residents with properties valued over 150,000 responded compared to 400 residents with properties values under 150,000. We will use a percent tabulation here to make a fair comparison between these two different sized groups. From the table below, it is clear that a majority of the residents are opposed to the new facility, and that property value seems to have only a slight influence.

	≥150,000	<150,000	total
in favor	12%	15%	14.4%
opposed	72%	60%	62.4%
no opinion	16%	25%	23.2%
totals	100%	100%	100%

Interpreting Graphs

Reading graphs is an important skill. From advertising product features to presenting new tax plans, industry and government are ever more frequently using graphs to convey information. In fact, the Graduate Record Examination usually contains several graph interpretation questions.

Example 2.3: Given the graph below answer the following questions. Source: *GRE 1993-4 Information and Registration Booklet.*

a) In which year did the number of graduate applicants increase the most from that of the previous year?

b) Which years showed decreases in the amount of applicants?

c) True or False:

i. The number of applicants more than doubled from 1982 to 1991.

ii. For each of the years 1983 to 1991, inclusive, the number of graduate student applicants was greater than that of the previous year.

iii. The greatest number of graduate students sent applications in 1990.

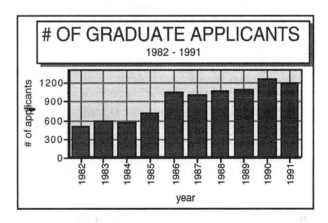

<u>Solution:</u> When reading a histogram, the height of the bar conveys the quantitative information. In this graph the height of the bar for any given year gives the number of graduate applicants for that year.

a) The largest jump is from 1985 (about 700) to 1986 (about 1050). So the answer is 1986.

b) 1984, 1987 and 1991 all are lower than their corresponding previous years.

c) i. true: From under 600 to about 1200.

ii. false: Several times it decreased, e.g., 1986 to 1987.
iii. true: 1990 has the tallest bar.

Example 2.4: Given the graph below answer the following questions. Source: *Barron's How to Prepare for the GRE General Test*, 8th Edition.

a) How many thousands of regular depositors did the bank have in 1975?

b) In which year(s) was there the greatest increase in the number of Christmas Club depositors over the previous year?

c) In 1974, what was the ratio of the number of Christmas Club depositors to the number of regular depositors?

d) Approximately how many more regular depositors were there than Christmas Club Depositors in 1981?

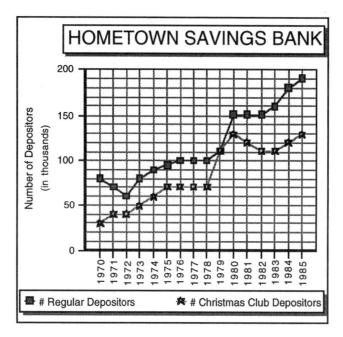

Solution: This is an example of a line graph. To read the number of deposits for a given year, start at the year and move vertically up until the appropriate line is reached, then move from there horizontally to left to estimate the number. Remember that the number of depositors given on the graph is "in thousands" and so should be multiplied by 1000 to get the actual number. This is a common technique for keeping scales easy to read.

a) Approximately 95,000 regular depositors.

b) 1979, since the increase from 70,000 in 1978 to 110,000 in 1979 was much more than in other years.

c) 60,000 to 90,000, since there were 60,000 Christmas Club depositors and 90,000 regular depositors.

d) 30,000, which is 150,000 (regular) - 120,000 (Christmas).

YOU TRY IT 2.4
Grades for an art quiz were distibuted as in the histogram on the right:

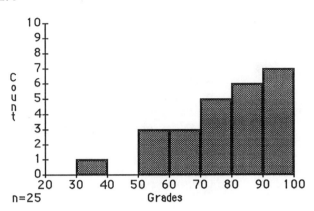

a) How many students took the quiz?
b) Make a frequency table for the scores on the quiz.
c) How many of the grades are 70 or better?
d) Draw a relative frequency histogram using the data from the frequency histogram.

YOU TRY IT 2.5

Answer the following questions based upon the graphs below from the Florence Dress Shop Sales and Earnings Report.   Source: *Barron's How to Prepare for the GRE, 8th Edition.*

a) What is the average in thousands of dollars of the sales for the period 1982 - 1985?

b) Which year showed the greatest increase in sales over the previous year?

c) What was the ratio of sales to earnings in 1980?

d) What were the total costs in 1982?

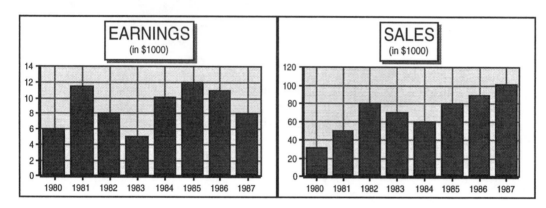

When Good Graphs Go Bad

Statistical graphs are very powerful tools for describing or summarizing information clearly and precisely. Unfortunately, graphs can also be used to mislead and confuse. A viewer must be alert to the misuses of graphs that result in poor communication of data, or even outright deception.

Inconsistencies in scaling are the most severe errors and can always be avoided. Scaling factors are often used inappropriately, as we see in this line graph of enrollment data:

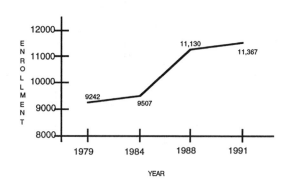

What's wrong here? Notice that the increments for the time scale on the horizontal axis are inconsistent. The hash marks are the same distance apart on the scale, yet they do not represent the same time periods! The visual impact leads the viewer to assume that each segment of growth took place over the same amount of time. This is obviously a false interpretation of the data. Note that the vertical and horizontal axes do not have to use the same scales, but each axis must have a consistent scale that is incremented evenly. Also remember that scales do not have to begin at zero.

YOU TRY IT 2.6

How could we correct the error in the graph above? Using the given data set, create a new line graph that would be a good representation of the enrollment data. Be sure that your axes are labeled clearly and scaled evenly.

Year	1979	1980	1981	1982	1983	1984	1985	1986
Size	9242	9794	9690	10051	9844	9507	9760	10419

Year	1987	1988	1989	1990	1991
Size	11070	11130	11501	11483	11367

Scaling that is consistent can still be used to create false impressions. **Exaggerating the scale** can change the emphasis of a change in the data. Look carefully at these three bar graphs of SAT data for ASU:

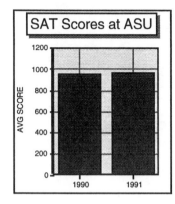 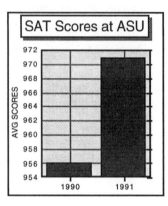 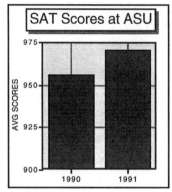

All three of these graphs have consistent, accurate scaling and are "honest" representations of the data, but they leave the viewer with very different impressions of the situation. The middle graph is greatly exaggerated and is more focused on impressing than on informing. The left hand graph has too large a scale, making it difficult to see any difference at all. The right hand graph gives a more reasonable view of the size of the increase in the scores.

Example 2.5: Look at the graphs below, and comment on the differences in scale. Which gives a clearer picture of the trend?

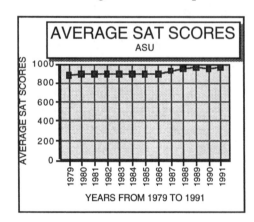 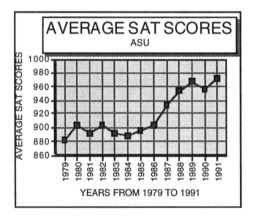

Solution: Here again, both of the graphs are scaled evenly on the vertical and the horizontal axes. The graph on the right has a finer scale than the graph on the left, but in this case the effect is positive. It is easier to see the subtle changes from year to year, and no particular year's data is over-dramatized.

So how do we choose an appropriate scale that is both accurate and fair? A graph can inform or mislead a viewer. The viewer must judge the author's intent: Is it to honestly represent the data, or is it to cover up faults and/or exaggerate successes? A good graph always presents data in a clear and unbiased form.

YOU TRY IT 2.7

Bob and Rob are twins about to graduate from the same university. Bob got off to a mediocre start his freshman year, but he soon got it all together and his grades drastically improved. Rob, on the other hand, started well but partied too much as the years went by. Here are the GPA stats on Bob and Rob:

	Freshman	Sophomore	Junior	Senior
Bob's GPA	2.5	3.0	3.4	3.5
Rob's GPA	3.3	2.9	2.5	2.2

Make two separate graphs for the GPA's of the two brothers. For Bob's, use a scale that exaggerates the improvement in performance over the four years. For Rob's graph, choose a scale that plays down the negative trend of his data.

Misuse of area or volume can also cause exaggeration of scale, especially in a **pictograph**. A pictograph uses symbols or symbolic pictures to communicate data patterns. Area and volume can often be used to trick the eye when making comparisons. What the eye sees and what the data actually says can be quite different.

Consider the picture below which is meant to show that if you invest $1 one year you will have $2 the next year by putting a $1 in a 1x1 square and a $2 in a 2x2 square.

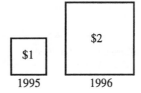

The proportions aren't right! The fault is misuse of area. Let's look closely:

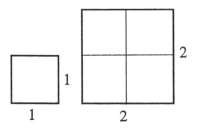

The area of the first square is 1 sq. unit (1 x 1 = 1), while the area of the second is 4 sq. units (2 x 2 = 4). The length of the side may be twice as long, but the area that you see is multiplied by four! When two- and three-dimensional representations are used in a statistical graph, areas and volumes should always be proportional to the data they represent.

As a rule, don't use this dimension deception; it can create an exaggerated impression of a change in the data.

Example 2.6: Comment on the use of area in this graph from *The Washington Post.,* October 25, 1978, page 1.

Solution: Certainly the purchasing power of the dollar has diminished, but is it as drastic as this pictograph would suggest? Not at all! The decrease in area of the dollars does not correspond to the decrease in buying power of the dollar. If fact, the 1978 dollar is only 14% of the area of the 1958 dollar, instead of the 44% it should be - a misuse of area.

The purpose of any graph is to speak clearly and communicate accurately: **Make the message clear.** Certainly the appeal of a graph, or the lack of it, is sometimes a matter of personal taste. Simplicity is much more powerful than clutter, however, when it comes to conveying information. Bad graphs can be found everywhere, even in popular newspapers. Consider the two graphs below, both of which incorporate data from the U.S. Census Bureau and Department of Labor on economic indicators.

The first graph leaves an overall impression of a decline in the inflation and unemployment rates, but it is difficult to interpret the actual numbers. Perspective is used in such a way that the scales are distorted - the higher values of the past are overemphasized while the current trend is diminished. This pictograph is designed to evoke an emotional response, not educate the reader on the actual numbers involved. When the picture dominates the data the true numbers can be easily lost, ignored or misunderstood.

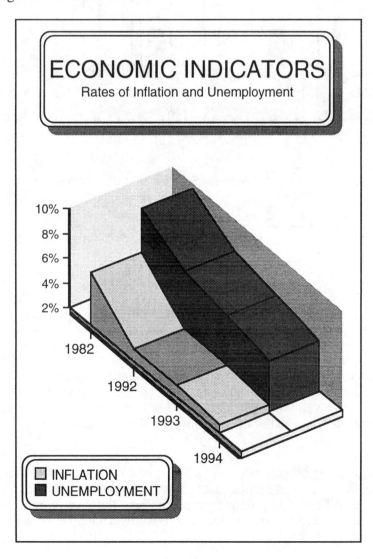

The second pictograph is rather extreme in its overemphasis on art. It makes an attractive illustration, but it doesn't do the best job of conveying specific information on the economic data. This graph also has too many things that it is trying to communicate, making the picture very busy. Few people reading the newspaper would really spend the time needed to accurately interpret the data in this graph.

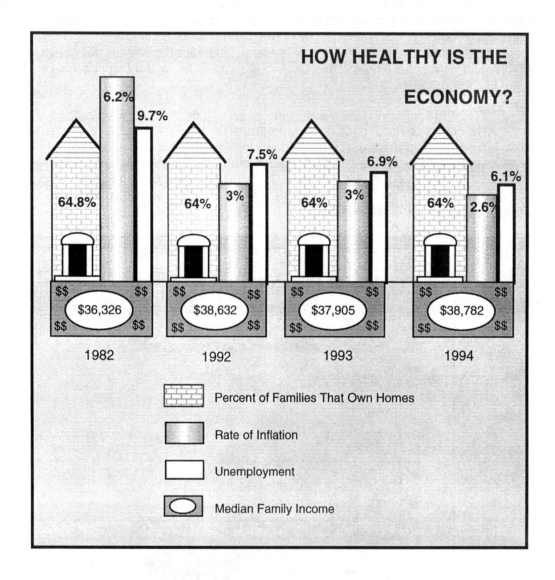

Pictographs can be visually very interesting and are quite effective when used to catch the reader's attention. However, as a mechanism for presenting data clearly and concisely, pictographs must be very carefully created, avoiding misuse of area, clutter and distorted scale, so as not to mislead the reader.

There are many other ways in which graphs can mislead and create false impressions. When creating graphs from data, care must be taken to avoid errors in presentation. When using graphs to interpret data, the viewer must be alert and discerning. Here are some basic guidelines to follow to help ensure that your graphs are good graphs.

Guidelines for Good Graphing
1) Keep it simple. Put major conclusions into graphical form. Don't try to say too many things through one graph.
2) Strive for clarity. Make the data stand out; avoid a "busy" picture.
3) Do not let pictures and labels interfere with the quantitative data on the graph.
4) Be certain that the scaling is consistent and labeled. Scale breaks may be used when necessary.
5) Avoid having too many (or too few) hash marks in the scale. A good rule of thumb is to have between 5 and 20, depending on the nature of the data.
6) Choose the range of hash marks to include the range of data. The data should fill up as much of the data region as possible.
7) Be creative, but be clear! Have a friend interpret your graph for you before you finish it. This is a good way to catch errors and misinterpretations.

YOU TRY IT 2.8

Discuss the good and bad features of this graph, similar to one from *USA Today*. Be sure to consider scale and clarity.

Drinking and driving

Alcohol involvement is the single largest factor in motor vehicle deaths and injuries, according to the U.S. Transportation Department. Meanwhile, arrests for drunken driving rose from 981 per 100,000 in 1980 to 1,048 in 1989.

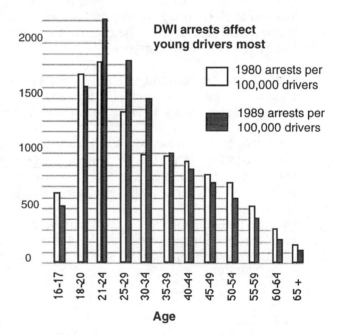

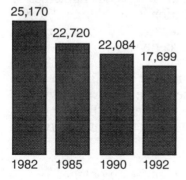

Alcohol-related motor vehicle deaths are down

Exercise Set 3.2

On the next page is a data set collected from 23 students who took this class at Appalachian State University in 1997. In problems 1 through 10, collect the corresponding data from your class, draw two graphs or tables, and compare your class to the given data.

1) Collect data on the number of siblings for each student. For comparison purposes, count all step- and half- siblings as well. Draw histograms for each set. The data for the given section are in column "SIBS." How does your class compare?

2) Collect data on the number of miles from home to campus for each student. For comparison purposes, round answers to the nearest 10 miles. Draw histograms for each set. The data for the given section are in column "DIST." How does your class compare?

3) Collect data on which class (freshman, sophomore, junior, senior) each student is in. Draw pie charts for each set. The data for the given section is in column "CLASS." How does your class compare?

4) Collect data on the number of seconds each student in your class can hold his or her breath. Draw histograms for each set. The data for the given section is in column "BREA." How does your class compare?

5) Collect data on how long students in your class think a minute is. To do this, time each student having them tell you when to start timing and then tell you to stop timing when a minute seems to have passed. Draw histograms for each set. The data for the given section is in column "MIN." How does your class compare?

6) Collect data on the number of siblings each student in your class has. For comparison purposes, count all step- and half- siblings as well. Draw histograms for each set. How does your class compare?

7) Collect data on gender from each student and on smoking. Generate cross tabulations for each set. The data for the given section is in columns "SEX" and "SMOKE." Is there a difference between the sexes? Did you expect a difference? Is your class different from the given section?

8) Collect data on gender from each student and on number of siblings (see problem 1). Generate cross tabulations for each set. The data for the given section is in columns "SEX" and "SIBS." Is there a difference between the sexes? Did you expect a difference? Is your class different from the given section?

9) Collect data on the number of siblings for each student (see problem 1). Draw Pie Charts for each set. The data for the given section are in column "SIBS." How does your class compare?

10) Collect data on the number of miles from home to campus for each student. Draw Pie Charts for each set with data grouped appropriately. The data for the given section are in column "DIST." How does your class compare?

ID	SEX	SMOKE	HEIGHT	ARM	FOOT	HEAD	SIBS	CLASS	DIST	MIN	REAL	NONS	BREA
1	M	Y	188	194	31	59.5	1	FR	200	65	12	7	58
2	M	Y	189	179	30	56	3	SO	180	57	7	3	50
3	M	Y	186	185.5	28	60	0	FR	40	61	9	5	51
4	M	Y	172	175	28	57	2	FR	60	51	5	2	59
5	M	N	177	175	29	59	0	SO	100	65	7	4	31
6	M	N	179.5	180	31	59	1	FR	110	68	4	3	35
7	M	N	190	188.5	32	61	4	FR	90	70	6	3	49
8	M	N	168	165	27	52	3	SO	90	56	9	2	62
9	F	Y	162	160	24	55	2	SO	120	55	7	6	35
10	F	Y	184	182	29	54.5	1	FR	250	60	5	4	15
11	F	Y	167	163	25	53	1	FR	310	42	6	3	32
12	F	N	157.5	157.5	24	54	0	FR	90	42	7	3	55
13	F	N	169.5	165	24	53	1	SO	110	120	8	3	20
14	F	N	164	156	25	55.5	1	FR	90	46	6	6	35
15	F	N	174	163	25	53	2	FR	140	58	10	2	65
16	F	N	159	152	25	54.5	1	JR	30	25	6	4	30
17	F	N	164.5	173	26	52	2	FR	200	55	7	5	51
18	F	N	169	174	28.5	56	1	FR	90	61	7	4	40
19	F	N	166	165	25	55	1	FR	70	64	11	2	66
20	F	N	176.5	171	26.5	55	1	FR	210	35	9	3	30
21	F	N	163.5	158	26	50.5	1	SO	100	42	11	3	57
22	F	N	161	146	25	51	3	FR	90	62	7	3	30
23	F	N	162	163	24	50	5	FR	50	45	8	7	37

11) A psychological study was done on the retention rate of reading material. (Source: Murgio's text *Communications Graphics*, 1969) The first graph is "without reinforcement" and the second graph is "with reinforcement." Reinforcement refers to additional study, rereading, homework applications, etc.

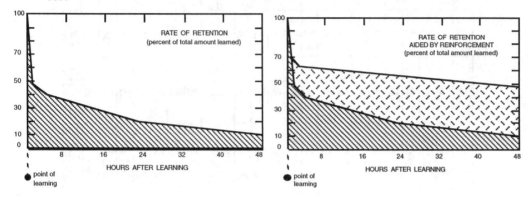

a) Without reinforcement, how much would the typical student remember after 8 hours? With reinforcement?

b) During what time interval do you see the most dramatic decrease in retention?

c) After two days, how much better is the retention rate for students with reinforcement than for those without?

12) Resident students at a university were asked to rate their compatibility with their roommates after one semester on a scale of 1 to 10, with 1 being low and 10 being high. The results are given in the graph below.

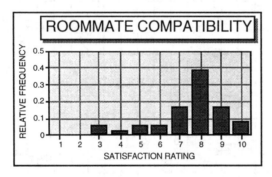

a) Approximately what percentage assigned a satisfaction rating of 7 to their roommates?

b) Approximately what percentage of the ratings were 7 or greater?

c) In general, would you conclude that these students were satisfied with their roommates?

13) The graph on the next page gives the fuel economy standards for the years 1978-1985. This graph is a reproduction of one from *The New York Times*, August 9, 1978, page D2.

a) What was the fuel economy standard set by Congress in 1982?

b) A line 0.5 inches long is used to represent the data value "18"; a line 4.8 inches long is used to represent the data value "27.5". What percent of 4.8 is 0.5? What percent of 27.5 is 18? What does this say about the use or misuse of perspective drawing in this graph?

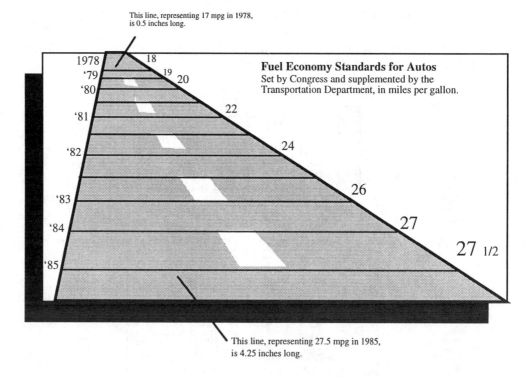

This line, representing 17 mpg in 1978, is 0.5 inches long.

Fuel Economy Standards for Autos
Set by Congress and supplemented by the
Transportation Department, in miles per gallon.

This line, representing 27.5 mpg in 1985,
is 4.25 inches long.

14) One of the pioneers of modern graphical design was William Playfair (1759-1823), an English political economist. The following graph is a reproduction of one published circa 1786 on British imports and exports to and from Denmark and Norway:

Exports and Imports to and from DENMARK & NORWAY from 1700 to 1780.

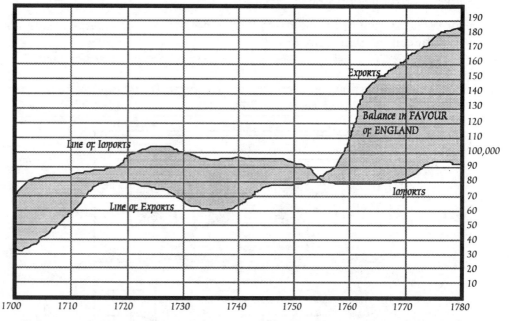

The Bottom Line is divided into Years, the Right hand line into L10,000 each.

a) In what year did British exports to Denmark and Norway begin to exceed imports?

b) Approximate the value (in British pounds) of exports for the year 1720.

c) Approximate the difference between the value of exports and imports for the year 1770.

d) How much did exports increase between 1740 and 1760?

15) Answer the questions below based upon the graphs. (Source: *Graduate Record Exam General Test*, 2nd Edition)

PERCENTAGE OF OCCUPATIONAL GROUPS VOTING REPUBLICAN

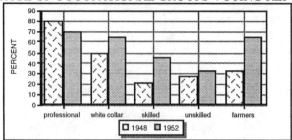

PERCENTAGE OF EDUCATIONAL GROUPS VOTING REPUBLICAN

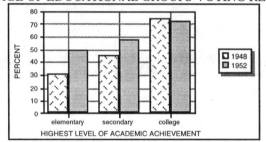

a) Which group showed the smallest percent increase of Republican voters between the elections of 1948 and 1952? (circle one)

elementary   secondary   skilled   unskilled   cannot be determined

b) Which group(s) showed an increase of almost 100% between the two presidential election years?

c) What percent of the elementary school-educated unskilled workers voted Republican in 1952? (circle one)

31%   50%   41%   81%   cannot be determined

16) Answer the questions below based upon the graphs. (Source: *Graduate Record Exam General Test*, 2nd Edition)

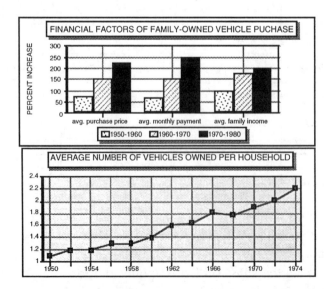

a) What was the difference in the percent increase in family income and the percent increase in the purchase price of a vehicle from 1960 to 1970?

b) Between 1950 and 1974, the average number of vehicles owned per household increased by approximately how much?

A fundamental knowledge of interpreting graphs is very helpful on the Graduate Record Examination (GRE), a test required for admission to many graduate schools. The questions below are similar in structure and content to those appearing in the quantitative section of the GRE. Questions 17 through 20 refer to the following graph.

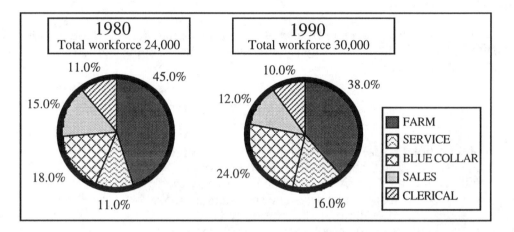

17) In 1980 there were how many thousands of farmers in the work force?
    (A) 9.9   (B) 10.8   (C) 15.1   (D) 24.0   (E) 45.0

18) What is the ratio of the <u>number</u> of blue collar workers in 1980 to the number of the same workers in 1990?
    (A) $\frac{3}{4}$ (B) $\frac{2}{5}$ (C) $\frac{3}{5}$ (D) $\frac{4}{7}$ (E) $\frac{5}{3}$

19) Approximately what was the percent increase in the <u>number</u> of blue collar workers in the workforce from 1980 to 1990?
    (A) 50%   (B) 70%   (C) 80%   (D) 90%   (E) 100%

20) From 1980 to 1990, there was an increase in <u>number</u> of workers in which categories?
    (A) All categories   (B) All except clerical
    (C) All except sales   (D) Service and bluecollar only
    (E) Service, clerical and bluecollar

3   SUMMARIZING DATA

We now begin the process of organizing and analyzing numerical data that have already been collected. We know how to create basic graphs for data, but for formal statistical analysis we'll want to quantify the characteristics of the distribution of the data. Are the data spread out or clustered near the center? Is the distribution lop-sided or symmetric? We will begin by describing summary techniques for **univariate** (one variable) data in this section and then describing relations between two variables in the next.

The summary statistics that we will examine in this section fall into three categories: shape, center, and spread. Before we can look at how to measure these characteristics for a data set, we need a tool for organizing the data.

Stem and Leaf Plots

One simple way to organize a small set of univariate data is to use a **stem-and-leaf plot**. This is a good way to order data values. Suppose grades on a recent math test were:

84	59	82	78	74	96	44	76	85	66
77	91	62	54	72	65	84	38	76	70

Each observation (grade) will be divided into a stem and a leaf. The tens digit will become the stem and the ones digit will become the leaf. A vertical line is drawn to separate the stems from the leaves. The stems are 3, 4, 5, 6, 7, 8, and 9, and after the first observation (84) is plotted, we would have the graph to the right. If the numbers are three digits, the leaves or the stems may be more than one digit long. If this happens put commas between them to separate.

```
3 |
4 |
5 |
6 |
7 |
8 | 4
9 |
```

The completed stem and leaf plot is then reordered by arranging all data in order (ascending or descending) from top to bottom, left to right. You can think of a stem-and-leaf plot as a histogram of the data that also displays the individual numbers.

```
3 | 8                          3 | 8
4 | 4                          4 | 4
5 | 9 4                        5 | 4 9
6 | 6 2 5                      6 | 2 5 6
7 | 8 4 6 7 2 6 0              7 | 0 2 4 6 6 7 8
8 | 4 2 5 4                    8 | 2 4 4 5
9 | 6 1                        9 | 1 6

    Stem and Leaf                  Ordered
                               Stem and Leaf
```

Once we have the data in order, we can start our analysis. We'll consider the shape of the distribution first.

The Shape of a Distribution

The distribution of a data set is said to be **symmetric** if the right and left halves (or upper and lower halves) are approximately mirror images of each other. The stem and leaf plot above has a distribution that is roughly symmetric. There are many other possible shapes. The ones we will give special names to here are:

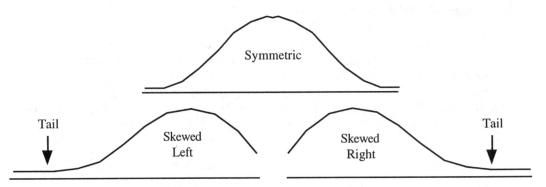

Symmetric

Tail

Skewed
Left

Skewed
Right

Tail

These are not the only possible shapes; for example, it is reasonable for a distribution to have more than one "hill." Recall from Section 1, however, that we expect data collected from a reasonably sized random sample to have a symmetric, **normal** distribution with the hill centered near the average value for the data. At this point we could quantify, i.e., assign a numerical value to, the amount of skew in a data set. Statisticians have a number of measures of skew (for example kurtosis) that we will not discuss here.

YOU TRY IT 3.1

Some students were asked to estimate their average monthly phone bill. Construct an ordered stem and leaf plot for the data. Describe the distribution.

| 10̶ | 55 | 14 | 28 | 15 | 18 | 40 | 38 | 80 | 15 |
| 75̶ | 22 | 65̶ | 35̶ | 20 | 24 | 12̶ | 45̶ | 25 | 43̶ |

10 12

The Center of a Distribution

We will discuss the two most common measures of center: Mean and median. The **mean** is the numerical value that locates the balance point or center of mass of the distribution. When given a random sample, we estimate the mean for the entire population by finding the **sample mean**, $\bar{x}$. If the sample consists of observations $x_1, x_2, x_3, ..., x_n$, then the sample mean is just an average of all observations in the sample. The mathematical formula can be written as:

$$\bar{x} = \frac{x_1 + x_2 + \cdots + x_n}{n} = \frac{\sum_{i=1}^{n} x_i}{n}.$$

Note the shorthand notation for the sum given above. The Greek letter sigma (Σ) indicates that what follows should be summed up with i replaced by the numbers 1 through n. Sometimes the i=1 below and the n above are omitted if it is clear what values of i to use. We'll see this mathematical shorthand again.

Example 3.1: Suppose 7 fish from Watauga Lake have the following lengths in inches:

$$4 \qquad 6 \qquad 7 \qquad 10 \qquad 12 \qquad 13 \qquad 18$$

Find the mean fish length for this sample.

Solution:
$$\bar{x} = \frac{4 + 6 + 7 + 10 + 12 + 13 + 18}{7} = 10$$

The **median** is the numerical value that divides the data in half. When given a random sample, we estimate the population median by the **sample median**, **M**. Assuming the sample is ordered, M is the middle observation if n is odd, and it is the average of the middle two observations if n is even. We can use a formula for the location of the median, loc(M) = (n+1)/2, or we can just look for the middle number.

Example 3.2: Let's go back to the 7 fish from Watauga Lake. Their lengths were:

$$4 \qquad 6 \qquad 7 \qquad 10 \qquad 12 \qquad 13 \qquad 18$$

Find the median fish length.

Solution: n=7 (odd) so there is a "middle" length and it is 10, so M = 10. If we use the location formula instead, we get loc(M) = 4, giving 10 for the value again.

Suppose another fish length is added to our data. Now their lengths are:

$$4 \qquad 6 \qquad 7 \qquad 10 \qquad 12 \qquad 13 \qquad 13 \qquad 18$$

Now n is even, and so there is no "middle" length. We take the mean of the middle two:

$$M = \frac{10 + 12}{2} = 11.$$

loc(M) = 9/2 = 4.5, telling us to take the average of the fourth and fifth numbers. Again we get 11.

Example 3.3: Let's look again at our ordered stem and leaf plot for Math 1010 test grades.

```
3 | 8
4 | 4
5 | 4  9
6 | 2  5  6
7 | 0  2  4 / 6  6  7  8
8 | 2  4  4  5
9 | 1  6
```

Calculate the mean and median, and discuss the shape of the distribution.

Solution: The mean is 71.65.

There are 20 (even number) data points so the median is located between the tenth and eleventh. To find the median, start at the top and count down 10 positions, or start at the bottom and count up 10 positions. If you start at the bottom, remember to count right to left as you go up. The median lies half-way between 74 and 76. Therefore, M = 75.

The data are very slightly more concentrated in the larger numbers, but not by much. These data are roughly symmetric. Turn the stem and leaf plot sideways to see that the tails are balanced.

Example 3.4: A professional basketball team has 6 players who earn $400,000 per year, 3 players who earn $1,000,000, and a superstar who earns $5,500,000. Compare the two measures of center for this data.

Solution: Let's calculate the mean first.

$$\bar{x} = \frac{6(400,000) + 3(1,000,000) + 5,500,000}{10} = 1,090,000$$

There are 10 players, so the median will fall between salary 5 and 6. Since the 5th and 6th data values are both $400,000, the median M = $400,000.

The superstar's salary has caused the mean to be much higher than the amount paid to most of the other players. The mean is very sensitive to just one or two extreme data values. The median will not change even if the superstar makes $10 million. The superstar's salary is just one of those that falls in the upper half of the data values.

The mean uses every data value in the sample, and so it is probably the most common measure of center. In the basketball example, however, the median gives us a more typical salary than does the mean. This does not say that the median should always be used when the data contain a few extreme outliers. Choosing a measure of center should depend on the intended use. You should ask yourself "Does the midpoint or the average best describe what I need to know?" To answer this question it is helpful to think in terms of balancing a teeter-totter.

- The mean is where to put the balance if the numbers are placed on the teeter totter in order, keeping track of how far apart the numbers are and stacking repeated values. Statisticians call this the best measure of center with respect to distance.

- The median is where to put the balance if the numbers are just placed in one row in order on the teeter totter, without worrying about how far apart they are. Statisticians call this the best measure of center with respect to location.

For the team owner in Example 3.4 the mean would be the best measure of center, since all the salaries must be paid. The median would be of more interest to the prospective player. In the same way, the mean value of the houses in a region would be the best measure of center for an insurance company to look at for calculating the cost of insurance, but a home buyer gets more information from the median price of houses.

YOU TRY IT 3.2

Suppose Company A and Company B both have 5 employees each. Find the mean and median salaries for each company. Which is the better measure of center? Which measure of center best typifies an employee salary? Which company would you rather work for? Here are the annual salaries for their respective employees:

Company A	Company B
$24,000	$36,000
$24,000	$36,000
$24,000	$36,000
$24,000	$36,000
$250,000	$36,000

mean 69,200   mean 36,000
med. 24,000   med. 36,000

The Spread of a Distribution

Measures of a distribution's shape and center are certainly useful, but they are not complete without some discussion of variability or **spread**. We need to know how the data are dispersed. We have said already that half the data values lie below the median and half the data lie above the median. Thus the **median** is the 50th percentile. Other important percentiles are the **lower quartile**, **Q1** (25th percentile) and the **upper quartile, Q3** (75th percentile). These are easily located, as we'll see on the next page. The extreme values we will denote by **Hi** and **Lo**.

The measures of spread that work best with the median are the **range** and the **interquartile range (IQR)**. The difference Hi - Lo is the range. The difference Q3 - Q1 is the IQR. The size of these ranges tells us how **variable** the data are. The median, quartiles, and extremes make up a **five number summary** for our data, a good way to look at shape, center and spread at a glance. Let's look at the set of estimated student phone bills given below, and learn about these numbers by doing a five number summary.

STUDENT PHONE BILLS

10	55	14	28	15	18	40	38	80	15
75	22	65	35	20	24	12	45	25	43

Begin by ordering the 20 data values in a stem and leaf plot:

```
1 | 0  2  4  5  5  8
2 | 0  2  4  5  8
3 | 5  8
4 | 0  3  5
5 | 5
6 | 5
7 | 5
8 | 0
```

To find the quartiles, think of Q1 as the middle of the data $\leq$ the median, and Q3 as the middle of the data $\geq$ the median. Now just use the same technique as for the median in the lower half of the data for Q1 and the upper half of the data for Q3.

M is the average of 25 and 28, so M = 26.5
Q1 is the average of 15 and 18, so Q1 = 16.5
Q3 is the average of 43 and 45, so Q3 is 44

Our 5 Number Summary is:

$$\text{Lo} = 10 \quad \text{Q1} = 16.5 \quad \text{M} = 26.5 \quad \text{Q3} = 44 \quad \text{Hi} = 80$$

We can also give formulas for the locations (*not* values) of the quartiles:

$$\text{loc(Q1)} = (\lfloor \text{loc(M)} \rfloor + 1)/2 \text{ and } \text{loc(Q3)} = n+1-\text{loc(Q1)}.$$

The $\lfloor \ \rfloor$ means to use only the whole number part of loc(M). Let's try:

loc(M) = (20+1)/2 = 10.5; average of the numbers in positions 10 and 11 = 26.5
loc(Q1) = (10+1)/2 = 5.5; average of the numbers in positions 5 and 6 = 16.5
loc(Q3) = (21 - 5.5) = 15.5; average of the numbers in positions 15 and 16 = 44.

Let's look at our measures of spread: Range = 80 - 10 = 70
                                          IQR = 44 - 16.5 = 27.5

YOU TRY IT 3.3

Use the phone bills data again from YTI 3.1. Assume all first row data are for males and all second row data are for females. Do five Number Summaries for each and list them side by side. Calculate the range and IQR.

Note: If the number of data points is odd, then the median will fall on a data value, (the single middle score.) When calculating the quartiles in this situation, use the median point as **both** a lower value and an upper value. If we look at the phone bill data without the $80 value, there are 19 data points, an odd number.

```
|--------------lower----------|
                             |------------upper---------------|
 10 12 14 15 15    18 20 22 24 25 28 35 38 40    43 45 55 65 75
                ↑              ↑                ↑
            Q1 = 16.5        M = 25          Q3 = 41.5
```

Box Plots

A box plot is a graph of a five-number summary. The box spans the quartiles, with an interior line marking the median. Whiskers extend out from this box to the extreme high and low data values. The plot gives a nice illustration of shape, center and spread when using measures related to the median.

The five-number summary for the estimated student phone bills is:

Lo = 10
Q1 = 16.5
M = 26.5
Q3 = 44
Hi = 80

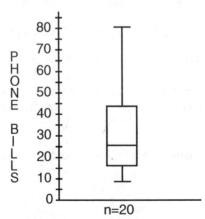

25% of all data lies between Lo and Q1 , 25% between Q1 and the median M, 25% between the median and Q3, and 25% between Q3 and Hi. So 50% of all data will lie in the box which spans the quartiles. The size of the box is the IQR.

Box plots are particularly useful for making comparisons between two subgroups. Look at the **side-by-side box plots** for male phone bills and female phone bills from the last You Try It:

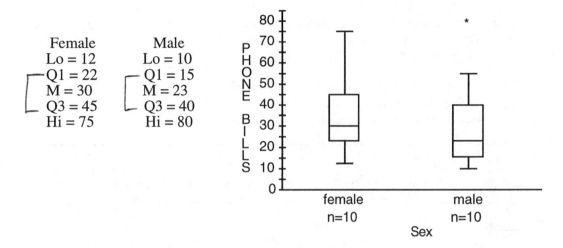

Female
Lo = 12
Q1 = 22
M = 30
Q3 = 45
Hi = 75

Male
Lo = 10
Q1 = 15
M = 23
Q3 = 40
Hi = 80

There is a value out of line with the rest of the values in the male phone bills, namely $80. Such an extreme value is called an **outlier**. In the box plot, the outlier has been starred and the "whisker" stops at the next highest number (55). This is usually the way an obvious outlier is pictured on a box plot. (If the outlier had occurred at the lower end of the range, it would be starred the same way and the whisker would stop at the next lowest number.) While we may designate an outlier with a special symbol, we never delete one unless we are certain that it is an error, or that there is some extenuating circumstance that makes it truly unlike the rest of the data.

Juxtaposing two box plots on the same scale allows us to make comparisons in a particularly efficient way. Let's look at the conclusions we can draw from this plot.

Example 3.5: Discuss the conclusions that are apparent in the box plots comparing male and female estimated phone bills.

Solution: In the phone bill box plots, we see that the middle 50% of the data for males is between $15 and $40. The middle 50% of the data for females is between $22 and $45. The median for the males was clearly lower than for the females. Overall, females have higher phone bills than do males. The highest phone bill, however, was that of a male. Male phone bills show more variability than those of females.

An Important Note on the Calculating Quartiles and Drawing Box Plots.

Quartiles and box plots are not standardized across different text books and software packages. The method described here is not the only technique used to calculate quartiles. You may find a textbook or statistical software package that doesn't include the median in the upper and lower halves of the data the way we do or doesn't average the values to get Q1 and Q3. This will change the values of the quartiles slightly, but usually not significantly.

You may also find that some sophisticated software packages and advanced textbooks on statistical analysis will use slightly different values called **Hinges** when drawing box plots. Hinges are calculated in a different manner from quartiles, but should be fairly close to the quartiles as we calculate them, at least for large data sets.

Also, some texts and software make a distinction between outliers that are "close" as opposed to those that are really extreme; so you may see box plots with different symbols for these two types of outliers.

In this text, we will always include the median in both halves of the data when calculating quartiles, and outliers will be subjectively chosen to be those data points that do not seem to fit the trend of the data.

YOU TRY IT 3.4

Below are boxplots for starting salaries of two different populations: first-year agriculture and engineering graduates of Tasmania State University.

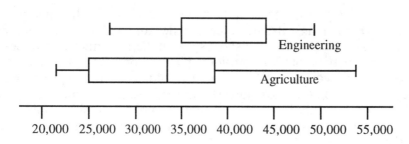

Starting Salary

a) The middle 50% of engineering salaries are between what two values?
b) The middle 50% of agriculture salaries are between what two values?
c) Which graduates are doing better overall?
d) At the very top levels which graduates are better paid?
e) What is the salary range for the bottom 25% of agriculture graduates?
f) Which salaries show more variability?

Box plots are not often seen in popular media, like newspapers, where more standard graphs like histograms and pie charts are prevalent. But for scholarly research in areas like the social sciences, box plots can convey a great deal of information in a concise, easy-to-read format.

Box plots give a nice picture of the shape, center, and spread of a data set when we use the median for the measure of center, but what if we want to use the mean? The appropriate measure of spread or variability in the data is the **standard deviation**. To introduce this idea, consider the individual "deviations from the mean." There is one for each observation in the sample, and it is calculated in the obvious way.

$$i^{th} \text{ deviation from the mean} = x_i - \bar{x}$$

Let's look at an example:

x	
1	The mean is $31/6 = 5.1667$
3	The first deviation is $1 - 5.1667 = -4.1667$
4	
6	The fifth deviation is $7 - 5.1667 = 1.8333$
7	
10	and so on...

To get the overall deviation we could average the deviations. Unfortunately, the sum of the deviations will always be zero; the negatives (ones below the mean) will total to the exact opposite in sign of the positives (ones above the mean). How can we counteract this "canceling out?" There are several ways to fix this problem, but the standard way is to sum the squares of the deviations. We'll compensate by taking the square root:

$$\sqrt{\frac{\text{sum of squared deviations}}{n}}$$

This still isn't quite right. For statistical reasons that we won't address here, when we use a *sample* to make a prediction about the whole population, we get a better estimate in we divide by n - 1 instead of n. So our actual formula for the standard deviation, s, is:

$$s = \sqrt{\frac{\text{sum of squared deviations}}{n-1}} = \sqrt{\frac{\sum (x - \bar{x})^2}{n-1}}$$

Notice the summation notation. Let's look at our example again:

x	$x - \bar{x}$	$(x - \bar{x})^2$
1	-4.1667	17.36139
3	-2.1667	4.694589
4	-1.1667	1.361189
6	.8333	.6943889
7	1.8333	3.360989
10	4.8333	23.36079

The sum of the squared deviations (the third column in the table) is 50.833333. Dividing by

$$n - 1 = 5$$

gives 10.16667. Finally, taking a square root gives:

$$s = 3.1885$$

What does the standard deviation tell us? It is most interesting when the data we are examining follow a normal (symmetric) distribution. Then roughly 68% of the data are within one standard deviation of the mean, and 95% of the data are within two standard deviations of the mean. This gives an excellent indication of how long the tails are in either direction:

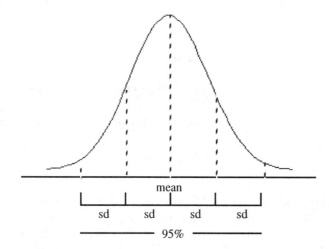

In this section we've seen how to describe the shape, center and spread of a distribution. These characteristics are effective summaries of the behavior of a data set, and form the basis of simple statistical analysis.

Exercise Set 3.3

1) The following table gives the American League leaders from 1970 to 1989.

Year	Player	Home Runs	Year	Player	Home Runs
1970	Frank Howard	44	1980	Reggie Jackson	41
1971	Bill Melton	33	1981	Four Players	22
1972	Dick Allen	37	1982	Thomas and Jackson	39
1973	Reggie Jackson	32	1983	Jim Rice	39
1974	Dick Allen	32	1984	Tony Armas	43
1975	Scott and Jackson	36	1985	Darrell Evans	40
1976	Graig Nettles	32	1986	Jesse Barfield	40
1977	Jim Rice	39	1987	Mark McGwire	49
1978	Jim Rice	46	1988	Jose Canseco	42
1979	Gorman Thomas	45	1989	Fred McGriff	36

Make a stem-and-leaf plot to show the distribution of league-leading home run totals. Are there any outliers? What do you think is the center for this distribution? 1981 was a strike-shortened season; what is the best way to deal with that data point?

2) An experiment was conducted to determine the effect of a drug on one's reaction time. The person is asked to depress a brake pedal whenever a light flashes. His reaction time for 12 trials (in milliseconds) follow. Make a stem-and-leaf plot of these observations. Are there any outliers or other unusual features?

96, 101, 100, 102, 138, 93, 99, 107, 93, 95, 100, 100

3) The following table gives the age at which U.S. presidents have died.

Washington	67	Fillmore	74	Roosevelt	60
Adams	90	Pierce	64	Taft	72
Jefferson	83	Buchanan	77	Wilson	67
Madison	85	Lincoln	56	Harding	57
Monroe	73	Johnson	66	Coolidge	60
Adams	80	Grant	63	Hoover	90
Jackson	78	Hayes	70	Roosevelt	63
Van Buren	79	Garfield	49	Truman	88
Harrison	68	Arthur	56	Eisenhower	78
Tyler	71	Cleveland	71	Kennedy	46
Polk	53	Harrison	67	Johnson	64
Taylor	65	McKinley	58	Nixon	81

Use this data to make a stem-and-leaf plot. What is the shape of the distribution?

4) Compute the mean and median for each of the following data sets. Note: The numbers are not sorted.

a) 4, 15, 2, 8, 4, 7, 10   b) 4, 2, 2, 6, 4, 4, 15, 8, 2, 17, 10, 4, 2, 6

c) 6, -3, 0, -11, 7, 110, -3

5) Another less common measure of center is the most frequently occuring number in the data, called the **mode**. The mode is rarely the best measure of center, although it is occasionally used. What are the mean, median, and mode for the data set in Exercise 3? Which of these values best represents a typical age at death for the presidents?

6) Discuss the pros and cons of choosing among the mean, median and mode (See exercise 5) as the measure of center for the following situations.

a) A local pediatrician wants to be able to tell new parents the typical waiting time for patients in her waiting room.

b) A hair-color manufacturer wants to add new colors to its line, and has collected data on preferred hair-color, the choices being 1 = blondes, 2 = reds, 3 = brunettes, 4 = blue-blacks.

c) A Realtor wants to advertise how inexpensive it is to live in a particular suburb.

7) a) Suppose you are interested in family incomes in a certain small town and you question some of the townspeople. Mr. Jones says, "the mean annual family income in the town is about $20,000." Mr. Adams say "Practically all of the families in the town have annual incomes less than $11,000." Is it possible that they are both being truthful? Explain.

b) Can the following statement be true? Less than 15% of the residents of Chicago earn more than the average wage for that city. Explain your answer.

8) According to the Department of Commerce, the mean and median prices of new houses sold in the United States in 1982 were $69,300 and $83,900. Which of these numbers is the mean, and which is the median? Explain.

9) Which of the mean, median, and mode (see Exercise 5) of a list of numbers must always appear as one of the numbers in the list?

10) Consider the data on home runs from Exercise 1.

   a) Construct the five number summary and a box plot.

   b) Comment on shape, center and spread.

   c) Based upon the box plot, where is the mean in relation to the median?

11) Consider the data on reaction time under the influence of drugs from Exercise 2.

   a) Construct the five number summary and a box plot.

   b) Comment on shape, center and spread.

   c) Based upon the box plot, where is the mean in relation to the median?

12) Consider the data on ages of death for U.S. presidents from Exercise 3.

   a) Construct the five number summary and a box plot.

   b) Comment on shape, center and spread.

   c) Based upon the box plot, where is the mean in relation to the median?

13) Twenty-five college students applying for graduate school take a personality test for admission purposes. Their scores are the following:

21	18	20	25	23	19	30	24	29	14	25	22	35
	26	23	16	33	25	22	17	34	22	31	27	25

   Construct a box plot. Comment on the shape, center and spread.

14) Below is a set of test scores for a class of twenty students:

   58 98 68 79 57 83 65 93 88 50 90 69 55 86 71 62 84 68 90 95

   a) The instructor would like to know the median test score and the range for the middle 50% of the scores. Create an appropriate graph for the data and answer these questions.

   b) Now the instructor would like to know the number of A's, B's, etc. The instructor uses a 10-point grading scale: 90-100 is an A, 80-89 is a B, 70-79 is a C, 60-69 is a D, and 0-59 is an F. Create an appropriate graph for the data and answer the question.

   c) Wouldn't you know it; now the instructor wants to know the percentage of students who received A's, B's, etc. Create an appropriate graph that will help answer the question, and give the answer.

15) Below is a list of data for 20 ASU students. The students were asked to report the number of hours of TV they watch in an average week.

Sex:	F	M	F	M	F	F	F	F	F	F
Hr. of TV/week:	6	14	20	1	16	4	35	10	2	35
Sex:	M	M	F	F	F	M	F	M	M	F
Hr. of TV/week:	2	10	15	5	7	2	20	10	3	9

Create a histogram for the data set as a whole. Use a class size of 5. Answer the following questions.

a) What is the shape of the distribution?

b) How many responses fell into the 11-15 class?

c) How many responses were less than 11?

d) In which class does the mean fall?

16) Using the data from Exercise 15, create side-by-side boxplots that compare males to females, and answer the following questions:

a) What is the range for males? for females?

b) What is the interquartile range for males? for females?

c) Based on the medians alone, which sex watches the most TV?

d) Using the information on the box plot to justify your response, decide which group is   more likely to watch more than 10 hours of TV per week.

17) Using the data from Exercise 15, create a relative frequency histogram for the data set as a whole. Use a class size of 5. Answer the following questions.

a) What percent of male responses were in the first class?

b) What percent of female responses were more than 10?

18) The Saxons and the Danes ruled England from 829 until 1066. The name of each ruler, the year his rule began, and the number of years he reigned are given below. Create a histogram for the years reigned data and answer the questions.

a)   What is the shape of the distribution?

b)   How many of the rulers reigned for less than 20 years?

c)   What was the longest reign? What was the shortest?

d)   What was the average reign?

Name	Year Reign Began	Years Reigned
Egbert	829	10
Ethelwulf	839	19
Ethelbald	858	2
Ethelbert	860	6
Ethelred I	866	5
Alfred	871	28
Edward I	899	25
Athelstan	924	16
Edmund I	940	6
Edred	946	9
Edwy	955	3
Edgar	959	17
Edward II	975	4
Ethelred II	978	37
Edmund II	1016	0
Canute	1016	19
Harold I	1035	5
Hardecanute	1040	2
Edward III	1042	24
Harold II	1066	0

19) Using the data from Exercise 18, create a box plot for the years reigned data and answer the questions.

a) What is the median for this data? Would you expect the mean to be above or below the median?

b) What is the IQR (InterQuartile Range)?

c) What percent of the rulers reigned between 2.5 and 6 years? Between 19 and 37 years? Between 0 and 19 years?

20) Through 1996, there have been 45 different Vice-Presidents of the United States, some of whom are still living. The names and the age of death for the deceased are:

Name	Age	Name	Age	Name	Age
John Adams	90	Andrew Johnson	66	Charles Curtis	76
Thomas Jefferson	83	Schuyler Colfax	62	John Garner	98
Aaron Burr	80	Henry Wilson	63	Henry Wallace	77
George Clinton	73	William Wheeler	68	Harry Truman	88
Elbridge Gerry	70	Chester Arthur	57	Alben Barkley	78
Daniel Tompkins	51	Thomas Hendricks	66	Richard Nixon	81
John Calhoun	68	Levi Morton	96	Lyndon Johnson	64
Martin Van Buren	79	Adlai Stevenson	78	Hubert Humphrey	66
Richard Johnson	70	Garret Hobart	55	Spiro Agnew	*
John Tyler	71	Theo. Roosevelt	60	Gerald Ford	*
George Dallas	72	Charles Fairbanks	66	Nel. Rockefeller	*
Millard Filmore	74	James Sherman	57	Walter Mondale	*
William King	67	Thomas Marshall	71	George Bush	*
John Brenkinridge	54	Calvin Coolidge	60	Dan Quayle	*
Hannible Hamlin	81	Charles Dawes	85	Al Gore	*

Create a histogram for the age of death data and answer the questions.

a) Would you say that the graph is skewed left, skewed right, or fairly symmetric?

b) How many Vice Presidents lived to be 80 years or older?

c) What was the shortest life span for any Vice President? What was the longest?

21) Using the data from Exercise 20, create a box plot for the age at death data and answer the questions.

a) What percentage of Vice Presidents died between the ages of 66 and 71? 80 and 98? 66 and 98?

b) Describe the overall distribution.

22) Here's some data on MPG from *Consumer Reports Magazine* (1992) for different sized cars.

Create a box plot for the entire data set and answer the questions.

a) What percent of cars get 24 mpg or better?

b) What is the best gas mileage of any car in the data set?

c) What is the range of the data?

d) Describe the shape of the distribution.

NAME	SIZE	MPG	NAME	SIZE	MPG
Acura Integra	Small	24	Subaru Legacy	Compact	23
Mazda Protege	Small	26	Chrysler LeBaron	Compact	23
Toyota Corolla	Small	26	Plymouth Acclaim	Compact	24
Geo Prizm	Small	26	Mazda 626	Compact	24
Saturn	Small	30	Ford Tempo	Compact	24
Mercury Tracer	Small	29	Nissan Stanza	Compact	21
Ford Escort	Small	34	Audi 80	Compact	22
Nissan Sentra	Small	35	VW Passat	Compact	24
Subaru Loyale	Small	26	Pontiac Sunbird	Compact	24
Hyndai Excel	Small	29	BMW 325i	Compact	21
Plym. Sundance	Small	28	Infiniti S45	Midsize	17
Isuzu	Small	27	Linc. Continental	Midsize	18
Mazda MX6	Sporty	25	Lexus ES300	Midsize	21
Dodge Stealth	Sporty	20	BMW 535i	Midsize	17
Toyota Celica	Sporty	27	Mitsub. Diamante	Midsize	20
Mazda MX5	Sporty	30	Acura Vigor	Midsize	23
Toyota MR2	Sporty	28	Toyota Camry	Midsize	21
Mitsub. Eclipse	Sporty	29	Nissan Maxima	Midsize	21
Hyundai Scoupe	Sporty	32	Ford Taurus	Midsize	21
VW Corrado	Sporty	25	Chry.New Yorker	Midsize	21
Isuzu Impulse	Sporty	29	Eagle Premier	Midsize	22
Geo Storm	Sporty	29	Olds Cutlass	Midsize	21
Mercury Capri	Sporty	28	Buick Regal	Midsize	20
Mitsub. Galant	Compact	24	Chevy Lumina	Midsize	22
Honda Accord EX	Compact	24	Lexus LS400	Midsize	19
Infiniti G20	Compact	24			

23) Using the data from Exercise 22, create side-by-side box plots that compare MPG for each size car. Answer the questions.

a) Which size car has the least variability in the data? Which size car has the most variability in the data?

b) For each size car, determine what percent get more than 24 mpg?

c) What is the IQR for small cars? What is the range for compact cars? What type of comparison statement could you make about these results?

24) An article is selected from a newspaper and the length of each word is recorded. There were 328 words in the article:

Word Length	1	2	3	4	5	6	7	8	9	10	11	12	13	14	15
Frequency	8	11	13	10	74	36	51	47	28	17	19	5	8	0	1

a) Create a histogram for the data and answer the questions.

   i. Would you say that the histogram is skewed left, skewed right, or fairly symmetric?

   ii. How many words were between 3 and 8 letters in length?

   iii. How many words were greater than or equal to 10 letters in length?

b) Create a relative frequency histogram for the data.

25) These data were collected by Nielson Media Research monitors for each 1991-1992 prime time television show. The measurements for each show's share (percentage of TV sets in use tuned to the show) and rating (each rating point represents 921,000 households).

a) Compare the data on Share for the three networks. Give the appropriate box plots. Based on the graph, which network does best? Worst? Explain.

b) Compare the data on Rating between the three networks. Give the appropriate box plots. Based on the graph, which network does best? Worst? Explain.

Show	Network	Rating	Share
60 Minutes	CBS	21.9	36
Roseanne	ABC	20.2	31
Murphy Browne	CBS	18.6	27
Cheers	NBC	17.6	27
Home Improvement	ABC	17.5	27
Designing Women	CBS	17.3	26
Coach	ABC	17.2	27
Full House	ABC	17.0	27
Unsolved Mysteries	NBC	16.9	27
Murder She Wrote	CBS	16.9	25
Major Dad	CBS	16.8	25
Monday Night Football	ABC	16.6	28
CBS Sunday Movie	CBS	15.9	25
Evening Shade	CBS	15.8	25
Northern Exposure	CBS	15.5	25
A Different World	NBC	15.2	24
The Cosby Show	NBC	14.8	24
Wings	NBC	14.6	23
Fresh Prince of Bel Air	NBC	14.5	23
America's Funniest Home Videos	ABC	14.5	22
20/20	ABC	14.4	26
Empty Nest	NBC	14.3	25
NBC Monday Night Movies	NBC	13.9	22
America's Funniest People	ABC	13.8	20
Family Matters	ABC	13.5	24
Rescue 911	CBS	13.5	22
LA Law	NBC	13.3	23
n the Heat of the Night	NBC	13.3	21
48 Hours	CBS	13.1	23
Golden Girls	NBC	13.1	23

LOOKING FOR LINEAR RELATIONSHIPS   4

All of the data analysis techniques we've considered so far assume that the data set is "univariate", ie., that there is only one variable for numerical data, perhaps with categories or names assigned. Often the goal is not to analyze one variable but to compare two. We refer to two variable data as **bivariate.** We will learn how to consider such questions as:

Does A predict B?   Does A cause B to occur?

Another question that arises in these situations is "how strong are these relationships?" Does the relationship account for 90% of the difference or only

50% ? This strength is referred to by statisticians as **correlation.** We will learn how to interpret correlation in this section as well.

Scatter Plots

If we want to explore data on the number of hours of exercise per week for a random sample of employees at a company, we are dealing with univariate data, or data in one variable, but if we throw in a second variable, like the number of sick days used per year, it becomes a bivariate situation. Is there any relationship between exercise and incidence of sickness? Let's take a look at the numbers from a random sample of 10 employees.

Hrs. of Exercise per week	12	5	0	7	3	4	6	6	2	4
No. of sick days per year	6	3	8	0	2	5	2	1	6	4

Consider each employee's individual data as an ordered pair on an x-y graph. The x-axis (horizontal) is the **independent** variable and the y-axis (vertical) is the **dependent** variable. In most cases, it is clear from the context of the question which variable is dependent on the other. In this situation we are trying to determine whether the number of days absent from work depends on the amount of regular exercise, so exercise is the independent variable and sick days is the dependent variable. Each ordered pair is now plotted on the graph. The first point plotted would be the point (12,6); the second point would be (5,3); the third point would be (0,8), and so on. Such a graph is called a **scatterplot.**

Example 4.1: Draw a scatter plot for the data given above.

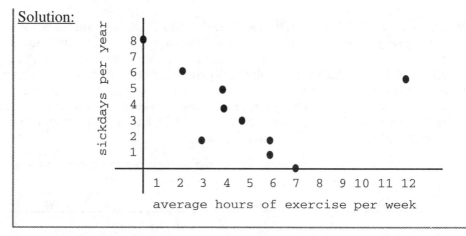

Do you see any trends in the data? Does there seem to be a relationship between exercise and sick days? If so, how strong is the relationship? Could we use this to make future predictions on employees' sick days? These are interesting questions, and statistics will help us answer them. Let's get started!

Outliers in Data

There is definitely a downward trend to the data: The more exercise, the less sick days used. Notice the position of the data point (12,6). This particular point seems to be unusual in the sense that it does not fit the pattern of the remaining data. Recall from our discussion in the previous section that such a point is termed an **outlier**. Outliers are often the result of human error, such as hitting a wrong key in entering data. However, an outlier may also indicate an exceptional circumstance. In this company, for example, there may be extenuating circumstances that would cause this person to use a large number of sick days in spite of significant exercise. Should outliers be considered when examining patterns in data? A general rule of thumb is this: If the outlier is clearly the result of human error, ignore it. If it is actual data include it in your data set.

Finding a Line That "Fits"

Once we have a scatter plot of the data, we will look for an obvious graphical relationship between the data sets. Most often we will be expecting a linear trend, but occasionally a curved relationship will appear. In the graph on sickdays versus exercise, it looks like we can model the relationship with a line, especially if we ignore the outlier. How do we find the line that fits a linear relationship?

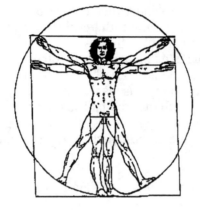

Example 4.2: ARE YOU A SQUARE? Leonardo DaVinci thought so! After studying the relationships between various dimensions of the human body, he suggested that our armspan is our height. It should come as no surprise that tall people tend to have longer arms than short people. You may readily observe this among your friends and classmates. However, the nature of the relationship between height and armspan may not be apparent. Investigate this relationship by plotting the data on 31 males between the ages of 18 and 24 below, and trying to use Davinci's idea that armspan = height.

	X	**Y**		**X**	**Y**
Person Number	Armspan (cm)	Height (cm)	Person Number	Armspan (cm)	Height (cm)
1	164	167	17	179	179
2	165	171	18	180	179
3	166	167	19	182	176
4	169	171	20	183	187
5	170	175	21	184	187
6	171	171	22	184	180
7	172	176	23	187	191
8	174	170	24	187	187
9	174	175	25	188	186
10	175	169	26	188	185
11	175	173	27	190	192
12	176	175	28	191	195
13	176	176	29	191	184
14	178	182	30	193	190
15	178	174	31	194	197
16	178	180			

<u>Solution:</u> Note that in the table, the corresponding heights are not in order, but as the armspans get longer, the heights tend to get longer as well. There are exceptions, such as person #5 being six cm taller than person #10 with an armspan that is five cm shorter. The height and armspan for most persons, however, are about the same.

The scatterplot given below helps us to see the trend in the data. The plotted points move upward as you read from left to right. When the armspans get larger, the corresponding heights tend to get larger.

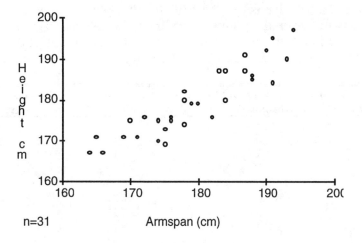

If Leonardo DaVinci was right, that is, if armspan (X) and height (Y) are the same for the general population, then we would expect the data to follow the pattern of the equation **height = armspan**, or Y= X. In the next plot, the line Y=X is drawn on the scatterplot. This gives us an impression of how much armspan and height differ for each individual. People with equal height and armspan have corresponding points that fall exactly on the line.

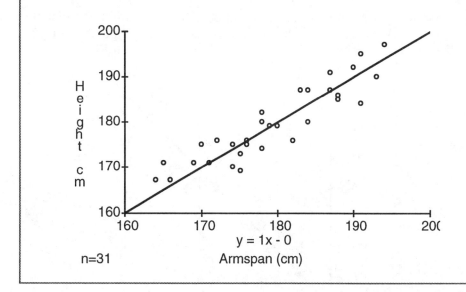

How close is the line to the points on the scatterplot? We want the line to be as close to the points as possible, so we add up all of the distances from the line to the points and try to make this sum as small as we can by moving the line slightly. The vertical distance between any particular point and the line can be computed:

$$|\text{(the y-value of the point)} - \text{(the y-value of the line)}| = \text{distance}$$

The absolute value assures us that all distances will be non-negative. If the point is on the line itself, the distance to the line will be zero. The data set below lists the distances between each of the 31 points and the line Y=X:

The smallest distance for a single piece of data is 0 and the largest is 7. Our goal is to find a line that is the "best fit" for the data. That is, we would like for the total distance from the points to the line to be as small as possible. The closer the line is to a point, the smaller the distance between them. If the data has a linear trend then the line will be "close" to all points, resulting in a smaller total.

Person Number	Distance from point to line in Absolute Value	Person Number	Distance from point to line in Absolute Value
1	3	16	2
2	6	17	0
3	1	18	1
4	2	19	6
5	5	20	4
6	0	21	3
7	6	22	4
8	4	23	4
9	1	24	0
10	6	25	2
11	2	26	3
12	1	27	2
13	0	28	4
14	4	29	7
15	4	30	3
		31	3
TOTAL VERTICAL DISTANCE			**93**

If the data set has no strong linear pattern, then the sum of the distances will be comparatively large. Consider how the last five distances in the table were computed:

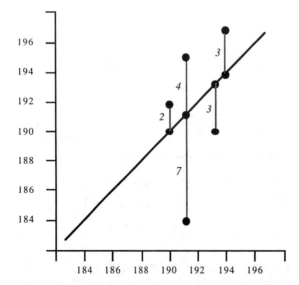

Person #27 corresponds to the point (194, 197) on the scatterplot. The point on the line is (194, 194), since the equation of the line is Y = X. If we subtract the y-values and take the absolute value, we get |197 - 194| = 3. Verify the results for a few other data points using this method.

Fitting a Line By Eye

An obvious question is "How do we come up with a line that models our data if there is no obvious one to use?" We could look at the scatter plot and try to fit a line "by eye", identifying a line that appears to be as close to the data as possible. For the height vs. armspan data we used the line Y=X, and this line approximated the data well. But if the equation of a line that fits is not obvious, how do we find it? All it takes is two well chosen points. Once we pick two points, we can write the equation for the line between them using algebra.

Example 4.3: The Bantam Sales Company wants to try to develop a testing method that will predict the first year sales for new salesmen. They gave a test to 10 new salesmen, recorded their scores and kept track of their sales performance for a year. Draw a scatterplot and fit a line by "eye".

Test Score	43	50	50	34	32	48	40	30	26	22
Sales (in $1000)	250	360	290	200	165	310	280	200	150	150

Solution: Here is a scatter plot of the data.

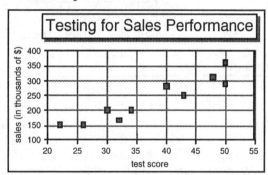

Now we can pick two points that appear to fit the data well. If you were to draw a line through the "center" of the scatterplot, which two points would you choose? Any reasonable choice is acceptable. How about the points (26, 150) and (48,310)? The graph below has this line drawn on it. Not bad! The points you choose do not have to be points in the data set itself; in fact, if you really want to minimize the distances between all the points and the line, the line that works best may not touch any of the points at all.

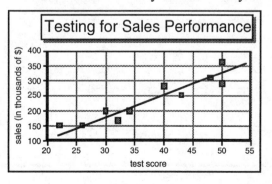

Writing the Equation for the Line

We need two things in order to write the equation for a line: the **slope, m,** of the line and a **point** on the line. If you have any two points, say (x_1,y_1) and (x_2,y_2), then the slope can be found from the formula

$$m = \frac{\text{change in y values}}{\text{change in x values}} = \frac{y_2 - y_1}{x_2 - x_1}.$$

If we apply this formula to (26,150) and (48,310), we get

$$m = \frac{310 - 150}{48 - 26} = \frac{160}{22} = \frac{80}{11}.$$

Avoid a common mistake: The order in which you subtract is important; if you used y_2 first on top use x_2 first on the bottom. Next we take the slope (m) and the point (x_1, y_1), and use the **point slope** formula

$$y - y_1 = m (x - x_1).$$

This will give us the equation of the line. Using the point (26, 150) and knowing that m = 80/11, the equation would be

$$y - 150 = \frac{80}{11}(x - 26)$$

$$y - 150 = \frac{80}{11}x - \frac{2080}{11}$$

$$y = 7.27x - 39.09 \quad \text{(rounding in the last step only!)}$$

YOU TRY IT 4.1

a) The point chosen for the formula above was (26, 150). However, it would not have mattered if we had used the point (48, 310) instead. Find the equation of the line using (48, 310) and verify that it is the same as the one above.
b) Write the equation for the line that connects the points (140,35) and (210,52).

Using Lines to Predict

One of the most common goals in creating these lines of best fit is to predict outcomes. Given a value of the independent variable, we can then predict what the resulting dependent variable would be.

Recall the goal of Bantam Sales Company managers in creating their test in the first place: It was to *predict* the sales performance of new salesmen. What would they predict for first year sales for Mr. Johnson, a new salesman who scored 35 on their test? The equation we have found,

$$y = 7.27x - 39.09,$$

is the equation of a line that fits the pattern of the data well. We can now use this equation to predict performance by simply substituting his test score (x) and determining his projected sales (y):

$$y = 7.27 (35) - 39.09 = 215.36$$

Since y was given in thousands of dollars, the projected sales for Mr. Johnson would be $215,360. Mr. Johnson could still turn out to be a lousy salesman or a real ace regardless of his test score. Based on the pattern of the remaining data, however, this is unlikely. The strength of our prediction is closely linked to the strength of the relationship of the data, as we will see later.

YOU TRY IT 4.2

Given the data below on the consumer price index (CPI) in the United States for 1970 (year 0) through 1980 (year 10), plot it and fit a line by eye. Use your line to predict the CPI for 1981 in two ways:

1) Use your graph to read the prediction.
2) Estimate the equation of your line and use the equation to make the prediction. The actual CPI for 1981 was 150.1. How close was your prediction?

Year	0	1	2	3	4	5	6	7	8	9	10
CPI	64	66.8	69	73.3	81.4	88.8	93.9	100	107.7	119.8	136

Regression Lines

If we are going to use lines to predict, we want to be sure that the linear models we are using are good ones. So far we have been fitting lines by eye. This method is quick and sometimes useful, but there is no guarantee that the line we choose will be the "best" line. Recall that we want our line to be such that the distances from the points to the line are as small as possible, but the line doesn't necessarily have to cross any point in the data. There are a variety of ways to come up with such equations through statistics, and the most popular method is called **least squares regression**. We show how to compute the equation for the regression line at the end of the section; for now we'll assume that you know how to use a computer spreadsheet or statistical software to find the equation.

The least squares regression line for the sales projection data above is

$$y = 6.768x - 18.295.$$

Using this equation, we would project sales for Mr. Johnson to be

$$y = 6.768 (35) - 18.295 = 218.585,$$

or \$218,585. Compare this to our previous prediction of \$215,360. The line of fit done by eye was good and comes close to the least squares estimate, but we'll see on the next page that the least squares regression line prediction allows more sophisticated statistical analysis.

Beware of Extrapolation

Predictions can be made for data using fitted lines such as the least squares regression line, or even a line fitted by eye. Caution must be taken, however, to ensure that these equations are not used to predict for unreasonable circumstances. Using a fitted line to predict the response of an independent (x) value outside the range of the data is called extrapolation. Extrapolation often produces unreliable results. Let's look at the Salesman Test Data again:

Test Score	43	50	50	34	32	48	40	30	26	22
Sales (in \$1000)	250	360	290	200	165	310	280	200	150	150

Notice that the test scores range from 22 to 50. What if we tried to predict the sales performance for a person scoring a 2 on the test? Using the regression line

$$y = 6.768x - 18.295,$$

the prediction would be
$$y = 6.768 (2) - 18.295 = -4.759.$$

This projects that the salesman scoring a 2 on the test would have a sales total of -\$ 4,759! This is an extreme example and obviously an impossible sales total, but this is the danger of extrapolation. If you move too far from the data in the data set, you are likely to get unreasonable or unreliable results. It is always best to use the predictor line to predict for numbers inside the range, or very close to the range, of the data set.

YOU TRY IT 4.3

Using the data and regression line just discussed, make a prediction for a salesman who scores a 95 on the sales performance test. Explain why this is an unreliable prediction.

 The Coefficient of Determination

How good is our least squares regression model at explaining the relationship between the two variables? We can expect data that have a linear trend will be better suited to a linear model than data that don't. Look at the plots below. We would expect that scatter plot A can be better estimated by a line than scatter plot B.

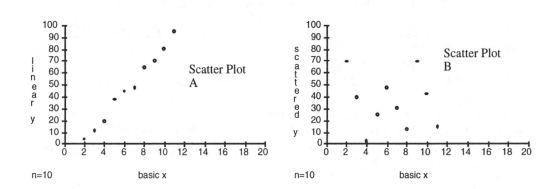

To answer the question "How linear is the trend in the data?", we turn to a special number r^2 called the **coefficient of determination**. This number is related to the proportion of the variation (or error) in the data that is explained by the regression line; in other words, r^2 gives a measure of the "% goodness of fit" of the regression line. The algebra involved in finding r^2 is discussed at the end of the section; fortunately, most computer programs that can find regression lines also give the value for r^2.

Clearly, some variables are more closely related than others. For example, you were probably not too surprised to see the relationship between armspan and height in an earlier example. It seems reasonable that taller people would have longer arms. But do you expect a correlation between arm span and GPA? Do you think people with longer arms perform better in school? Probably not! What the coefficient of determination r^2 gives us is a single number which measures the **strength** of the *evidence* of a relationship between two variables. Let's look at some of the properties of r^2 :

1) $r^2 \leq 1$. This is equivalent to saying that $0 \leq r^2 \leq 1$.

2) r^2 measures the proportion of variation in the dependent variable that is *explained* by the linear model. r^2 is usually reported as a percent; the closer it is to 100% the better the line explains the relationship between the two variables.

4) $r^2 = 1$ indicates a perfectly linear association in the data. If $r^2 = 0$, then the data have no linear association at all. The *evidence* for the linear relationship is considered to be:

strong: $65\% < r^2 \leq 100\%$,
moderate: $25\% < r^2 \leq 65\%$,
weak: $10\% < r^2 \leq 25\%$.

The symbol r^2 is used because the coefficient of determination is the square of another number sometimes used by statisticians, called the **Pearson Correlation Coefficient** and denoted **r**. r^2 is the preferred measurement of the strength of a relationship, but sometimes **r** will be reported instead. The only advantage of using **r** is that the sign is chosen to indicate the direction of the relationship. If **r** is positive, the data have an upward (increasing) pattern, i.e., as x gets larger, y tends to get larger. If **r** is negative, the data have a downward (decreasing) pattern, i.e., as x gets larger, y gets smaller. Remember:

When assessing the strength of a relationship use r^2.

In scatter plot A above, the value of the correlation coefficient **r** is 0.99. The fact that **r** is positive suggests an upward trend in the data. Since $r^2 = 98\%$ is very close to 100%, we say that there is evidence for a strong linear association between the variables. Both of these things are clearly seen on the scatter plot.

A contrast can be seen in scatter plot B, where **r** = -0.15. The negative property of **r** suggests a downward trend, but the proximity of $r^2 = 2\%$ to 0 suggests that there is no evidence of a relationship between these variables. This is evident from the graph, also.

Example 4.4: Using a statistics package on the height (y) versus armspan (x) data given in the beginning of the section, we find that the regression line and correlation coefficient are:
$$y = 0.898 x + 18.4$$
$$r = 0.912$$

Discuss how well the line fits the data, and estimate the height of someone with an armspan of 167 cm.

Solution: First, we must state that it makes sense that height and armspan might be related.

r = 0.912, implying an increasing linear relationship between arm span and height.

$r^2 = (0.912)^2 = 0.832$ or about 83%. This says that there is strong evidence to support our linear model for the relationship between arm span and height, and we can predict about 83% of the variation in height for people.

A person with arm span = 167 has a predicted height

$$y = 0.898 \ (167) + 18.4 = 168 \ cm.$$

Sometimes two data sets will have a high correlation but the relationship is contrived. Suppose someone came to you with the following analysis: "I have compared the data on total rainfall in the Australian Outback (in inches per year) with data on the size of the corn crop in Iowa (in bushels harvested per year) for the last 30 years and found a correlation with $r^2 = 94\%$. Therefore, the amount of rain falling in Australia directly effects the corn crop in Iowa!" While the correlation coefficient might be correctly calculated, the relationship must be coincidental. Use your common sense.

Example 4.5: Consider this data which comes from the final statistics of the 1991-92 ASU Men's Basketball Team. For each player, the number of fouls and the number of points scored for the season is recorded and put into a scatterplot.

Player:	#1	#2	#3	#4	#5	#6	#7	#8	#9	#10	#11	#12
Fouls:	85	81	59	72	44	76	54	39	4	4	7	23
Points:	462	415	298	300	234	220	133	107	19	19	7	16

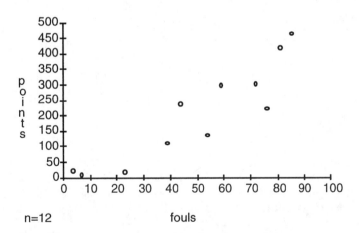

n=12

The correlation coefficient **r** here has a value of 0.92, giving $r^2 = 84\%$ which suggests a strong association between the two variables. But do you really think fouls are directly responsible for the number of points scored, or could there possibly be some other *confounding* factor that causes these two variables to behave this way?

Solution: It would seem that if a player commits a lot of fouls, he would spend some time on the bench having fouled out. You can't score points if you're not in the game!

There is another plausible cause for the increasing trend that we see in both variables here, and that is the number of minutes played per game. If a player gets a lot of playing time, it is expected that he will get a lot of fouls . . . and a lot of points. A measure of a player's aggressiveness might also be another related factor. A very aggressive player is more likely to commit fouls and score points than a more passive player.

A high degree of correlation, then, does not automatically imply that one variable directly "causes" another variable to behave the way it does. Interpreting statistical graphs without applying common sense to the situation can lead to illogical and incorrect conclusions about the data. A healthy dose of reasoning should always accompany statistical interpretations.

YOU TRY IT 4.4

Given the data below on the consumer price index (CPI) in the United States for 1970 (year 0) through 1980 (year 10), draw a scatter plot, find and plot the regression line, and find the coefficient of determination. (This is the same data you fit a line to by eye in YTI 4.2.)

a) Explain how well the linear regression model fits the data.

b) The actual CPI for 1981 was 150.1. How well does the regression line predict this?

Year	0	1	2	3	4	5	6	7	8	9	10
CPI	64	66.8	69	73.3	81.4	88.8	93.9	100	107.7	119.8	136

Extra for Experts

Linear regression is a very important tool in data analysis. It can be helpful when using such a tool to have an understanding of how it works. Let's look at the procedure involved in fitting a line $Y = Ax + B$ to pairs of data.

Given a bivariate data set containing n (x,y) pairs, we want to find values for A and B in the formula $Ax + B$ so that we have the smallest "squared error" between the line and the points. Mathematically, we want to minimize

$$[y_1 - (Ax_1 + B)]^2 + [y_2 - (Ax_2 + B)]^2 + [y_3 - (Ax_3 + B)]^2 + \ldots + [y_n - (Ax_n + B)]^2$$

$$= \sum (y - (Ax + B))^2$$

The formulas for A and B are given below. Deriving them uses techniques from calculus which we will not discuss here.

$$A = \frac{n \sum xy - \sum x \sum y}{n \sum x^2 - (\sum x)^2} \qquad B = \bar{y} - A\bar{x}$$

Notice that these formulas require values for all the various sums (Σ) and means involved. Let's use an example to illustrate how to efficiently calculate them. Calculate the equation of the least squares line for the data:

$$(5, 2), (7, 5), (9, 9), (11, 17)$$

Let's organize our information in a table.

n = 4	x	y	x^2	xy	y^2
	5	2	25	10	4
	7	5	49	35	25
	9	9	81	81	81
	11	17	121	187	289
Σ:	32	33	276	313	399
mean:	8	8.25			

Notice that we have made a column for the first coordinate (x), the second coordinate (y) and all the other combinations needed for the formulas. Then at the bottom of each column, we have calculated the sum of the column, and the mean if we need it. Now we can find A and B:

$$A = \frac{4(313) - (32)(33)}{4(276) - (32)^2} = 2.45 \qquad B = 8.25 - 2.45(8) = -11.35$$

Thus the regression line is $Y = 2.45x - 11.35$.

What about the coefficient of determination? It is a measure of how much of the "error" or uncertainty in using the least squares line is from the line itself rather than being from the original data. We take the "error" in the line (i.e., the squared distances from each point to the line) and divide by the "error" in the data (i.e., how far away from the mean the data stray,) to get a "percent badness of fit". To get a "percent goodness of fit", subtract this from 100%!

Error in the line: $\sum [y - (Ax + B)]^2 = \sum y^2 - B \sum y - A \sum xy$

Error in the data:
$$\sum (y - \bar{y})^2 = \sum y - \frac{(\sum y)^2}{n}$$

Using the (easier to calculate) second formulas for each gives a percent goodness of fit of

$$r^2 = 1 - \frac{\sum y^2 - B \sum y - A \sum xy}{\sum y^2 - \frac{(\sum y)^2}{n}}$$

This is actually the square of the correlation coefficient, r, discussed in this section, where we choose + if the data have an upward trend and - if the data have a downward trend.

$$r = \pm \sqrt{\text{goodness of fit}}$$

In our example $r^2 = 0.947 \approx 95\%$, giving $r = +0.973$, since the data has an upward trend.

Summing It Up

Numerical information is all around us; statistics gives us a way to interpret these data and to make intelligent decisions based on it. We have also seen how statistics can be misused, resulting in false impressions and misinterpretation of the facts. When used correctly, however, statistics gives us the ability to communicate large amounts of information clearly and to make reasonable predictions.

Exercise Set 3.4

Round answers as specified in each problem. If no rounding rule is given, round answers to the least accurate measure in the problem:

For Exercises 1-6, round to one decimal place of accuracy:

1) Find the equation of the line between the two given points:
   a) (9,50) and (2,78)   b) (1.5, 3.4) and (6.8, 19.2)

2) Find the equation of the line between the two given points:
   a) (1100,200) and (1750,340)   b) (65.1, 23.4) and (90.0, 15.6)

3) Find the equation of the line between the two given points:
   a) (2,6) and (5,7)   b) (1759 , 2564) and (1833, 1040)

4) Find the equation of the line between the two given points:
   a) (.3, .1) and (.9, .6)   b) (140, 5) and (620, 15)

5) Find the equation of the line between the two given points:
   a) (110, 60) and (175, 32)   b) (6, 93) and (12.5, 110)

6) Find the equation of the line between the two given points:
    a) (16, 78) and (30, 115)   b) (.8, 2.5) and (4.3, 7)

7) The following data set compares a child's age with the number of
    gymnastics activities he or she was able to complete successfully:

Age:	2	3	4	5	6	7
No. of activities:	5	5	6	10	9	11

    Below are two scatterplots of the same data, each with a different equation
    for the line of fit. Determine which equation has the best fit by finding the
    sum of the vertical distances from the points to the line for each. Which line
    do you think is closest to the least squares regression line? Explain your
    choice.

    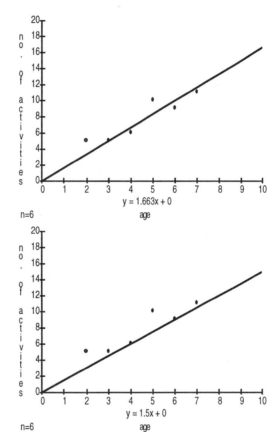

8) An automobile dealer records the number of cars sold and the prime interest
    rate for six 2-month periods.

Prime rate (%):	15	14	13	12	11	10
Sales (no. autos):	116	132	148	136	155	184

    Below are two scatterplots of the same data with different equations for the
    line of fit. Determine which equation has the best fit by finding the sum of
    the vertical distances from the points to the line for each. Which line do you
    think is closest to the least squares regression line? Explain your choice.

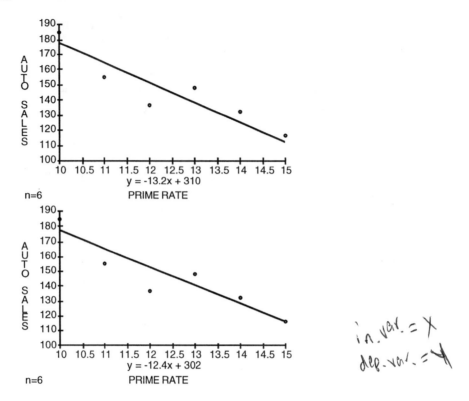

n=6   PRIME RATE
y = -13.2x + 310

n=6   PRIME RATE
y = -12.4x + 302

in. var. = X
dep. var. = Y

9)   For each situation given, which is the independent variable x and which is
      the dependent variable y?

      a) A medical study wants to determine if pulse rates are affected by body
      temperature.

      b) A burglar robs a home, leaving only his footprints behind. In trying to
      determine more information in these circumstances, the police department
      conducts a study to see if foot length can be used to predict height.

10)   For each situation given, which is the independent variable and which is the
      dependent variable?

      a) A ski slope manager collects data to determine the effect of natural
      snowfall on the number of lift ticket sales.

      b) Do smaller dogs live longer lives? The Humane Society would like to
      know if a dog's life span is related to its weight.

11)   A study was conducted to determine the effects of sleep deprivation on
      subjects' ability to solve simple problems. The following results were
      obtained:

Hours Without Sleep:	8	10	12	14	16	17	20	22	24	26
Number of Errors:	8	6	6	10	8	14	14	12	16	12

      What are the independent and dependent variables here? Create a scatterplot
      of the data. Use the points (12,8) and (23,13) to fit a line; Find the equation
      of this line of fit, rounding to the nearest hundredth. Then use it to predict
      the number of errors for someone who goes 15 hours without sleep.

12)   Midterm grades, along with the number of absences, were posted for a
      particular math class of 25 students. The results are listed below:

Midterm average	74	11	38	75	82	80	48	83	39	79	85	65	37
No. of absences	9	14	8	12	4	1	2	2	13	0	4	9	12
Midterm average	67	85	81	60	92	80	76	74	36	75	88	63	
No. of absences	1	1	4	4	0	2	1	2	14	6	0	6	

a) What are the independent and dependent variables here? (Use your common sense, not the order from the chart.) Create a scatterplot of the data and fit a line by eye. Find the equation of your fitted line and use it to predict the average for a student who misses 10 days.

b) The least squares regression line for the scatterplot is found to be

$$y = -3.235x + 83.763$$

Use this line to predict the average for a student who misses 10 days. How close to this value was your estimate from part a?

c) The value of the correlation constant for this data is $r = -.75$. Calculate and interpret r^2. Explain what this means about the strength of the relationship between absence and performance.

13) Match the graphs with the appropriate correlation coefficients:

    1) $r = -0.48$ 2) $r = 0.36$ 3) $r = 0.81$ 4) $r = -0.72$

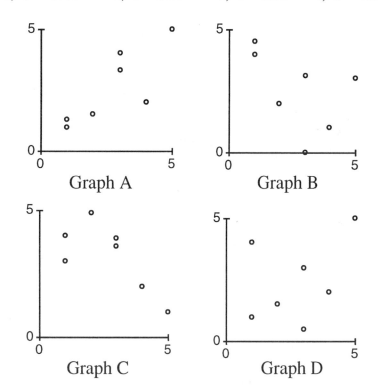

Graph A   Graph B

Graph C   Graph D

14) Mark Twain in *Life on the Mississippi* discussed how the river had gotten shorter over the years as segments of the river straightened themselves out. (See excerpt in Exercise Set 4.1). He gave the following as the length (in miles) of the Mississippi River:

Year:	1700	1720	1875
Length:	1215	1180	975

a) Graph these points and fit a line through them as best you can. Then make a prediction about the present length of the Mississippi River.

b) He predicts that the river will have length 0 at some point in the future. When will that be, according to your equation?

c) How accurate do you think these predictions are? Why?

15) A study comparing the number of years of experience of New York City patrolmen to the number of tickets they give per week revealed the following data:

```
No. years experience  3   8   2  15   5  20   1  10   7  12
No. tickets per week  42  30  54  12  32   8  75  28  20  15
```

a) Create a scatterplot for the data and fit a line by eye; find the equation for the fitted line and use it to predict the number of tickets per week for a patrolman with 6 years of experience.

b) The regression line for this data is $y = -2.932x + 55.038$. What does this line predict for a patrolman with 25 years experience? Explain why this prediction is not reliable.

c) The value for the correlation coefficient here is $r = -0.86$. Calculate and interpret r^2. Explain what this says about the strength of the relationship between years on the job and number of tickets issued.

16) 86 students were asked to estimate the number of calories in a variety of foods. Given below are the results along with the scatterplot for the data.

Food	True Calories	Mean Estimated Calories
1 med. banana	100	76
2/3 cup broccoli	29	60
1 oz. semi-sweet chocolate	147	202
1 serving baked flounder	204	165
1 med baked potato	155	128
6 vanilla wafers	120	254
2 oz. pork link sausage	125	195
1 7" diameter waffle	210	274
4 lg. fried shrimp	259	60
1 cup spaghetti	155	147
3 oz. rib roast	375	344

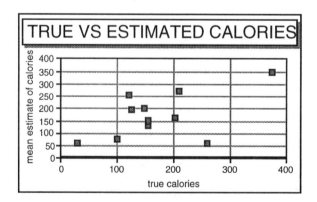

a) Based on the scatterplot, would you say that calories are more likely to be overestimated or underestimated? How did you decide?

b) Fit a line by eye to the data and write the equation. Use your equation to predict the estimated calories for a food containing 300 calories.

c) The least squares regression line for this data is $y = 0.56x + 76.75$. Use this equation to predict the estimated calories for a food containing 300 calories. How close are your estimates from parts b and c?

d) The correlation constant here is r = .56. What does this say about the strength of the relationship between true and estimated calories?

17) Here are some statistics from the college football bowl games played on January 1, 1962. Make a scatterplot of the data, and find the equation of a line fitted by eye. Discuss the importance of ball control by measuring the relationship between each team's number of first downs and its score in the game. What would your line predict for a team who had 20 first downs?

Bowl	Team	First Downs	Score
Rose	Minnesota	21	21
	U.C.L.A.	8	3
Sugar	Alabama	12	10
	Arkansas	7	3
Cotton	Texas	12	12
	Mississippi	17	7
Orange	L.S.U.	19	25
	Colorado	7	7

18) In the Height vs. Armspan example (ARE YOU A SQUARE?) in the text, we looked at claim that height and armspan are the same. The least squares regression line for the data is y = 0.9x + 18.38 with a correlation coefficient of r = 0.91.

a) What would this line predict as the height for someone with an armspan of 184 cm? ...195 cm? ...169 cm?

b) Based on the given regression line, would you conclude that height and armspan could be equal for a given individual? Is this true of the general population? Explain.

c) What does the correlation coefficient suggest about the strength of the relationship between armspan and height?

   The following exercises require computation of the regression line.

19) Many manatees are killed each year by power boats operating in the coastal waters of Florida. The game commission is concerned about the decline in manatee population and wants to determine if there is a relationship between the number of power boat registrations and the number of manatees killed each year. It found the following data. (Source: *For All Practical Purposes*, COMAP, 1994)

a) Which is the independent variable and which is the dependent variable?

b) Plot the data and point out any outliers. Find the regression line and use it to predict the number of manatees killed in a year if the number of registrations increases to 700,000.

c) What is the strength of the relationship for these data? Should the game commission be concerned about the number of registrations if it wants to save the manatees? Explain.

year	boats registered	manatees killed
1977	447,000	13
1978	460,000	21
1979	481,000	24
1980	498,000	16
1981	513,000	24
1982	512,000	20
1983	526,000	15
1984	559,000	34
1985	585,000	33
1986	614,000	33
1987	645,000	39

20) The table below gives data on the world records in the 800-meter run for men and women in this century. Since the data is given in 10-year intervals, the years have been coded 1-9 for convenience. (This is the "x" in the table.)

a) Create two scatterplots of time versus performance - one for men and one for women.

b) What is the value of the coefficient of determination for men? for women? What does this suggest about world records in the 800-meter run over time? What is the regression line for men? for women? What would these lines predict as world records in 1996?

c) The president of the New York Road Runners Club, Fred Lebow, is quoted as saying "Women will never pass men. Never, never." According to your lines when will the women catch up to the men's world record? If the downward trend in the records for women continues to hold, when will the women run the 800-meter in 0 time? How certain do you feel about your conclusions? Explain.

Men's and Women's World Records in the 800-Meter Run

Year	x	Men's Record	Women's Record
1905	1	113.4	---
1915	2	111.9	---
1925	3	111.9	144.0
1935	4	109.7	135.6
1945	5	106.6	132.0
1955	6	105.7	125.0
1965	7	104.3	118.0
1975	8	104.1	117.5
1985	9	101.73	113.3
1994	9.9	101.73	113.3

5   NOT EVERYTHING IS LINEAR

There are many data sets for which fitting lines gives an excellent indication of the relationship between two variables. But sometimes lines are not appropriate. In this section our goal is to make you aware that non-linear relationships exist, and that there is a myriad of choices for possible curves when dealing with non-linear data.

Consider the following data set taken from 15 males of various ages.

ID	Height(m)	Armspan (m)	Weight (kg)
1	0.75	0.71	12
2	0.86	0.86	12
3	0.95	0.94	14
4	1.08	1.09	17
5	1.12	1.11	22
6	1.26	1.26	25
7	1.35	1.35	35
8	1.51	1.50	41
9	1.55	1.53	46
10	1.60	1.62	50
11	1.63	1.62	51
12	1.67	1.65	54
13	1.71	1.73	65
14	1.78	1.81	72
15	1.85	1.88	88

Here is a plot of height versus arm span, which as we have seen before for other data, is highly linear:

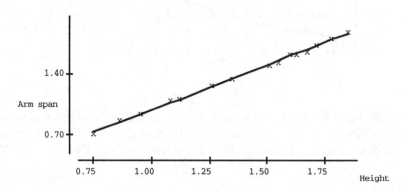

The regression information is

$$\text{Arm span} = -0.0424 + 1.03 \times \text{Height} \quad \text{with} \quad r^2 = 99.8\%,$$

which reinforces our belief that these data are related linearly. Now consider the graph of height versus weight:

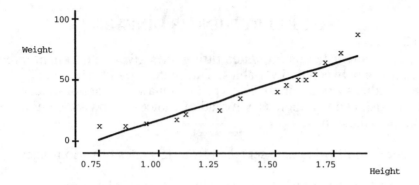

The r^2 here is not bad at 88%, but the relationship between the line and the data points is interesting. Notice that the lowest height data points are above the line, then as we move to the right the data drops below for a while and then rises above the line again at the end. That isn't what happens with the arm span versus height data; in that graph, the data points don't have any "above versus below the line" pattern, or any other discernable pattern for that matter.

This sort of pattern in the points compared to the line is indicative of a curved, non-linear relationship. Here is the same height and weight data with a curve superimposed:

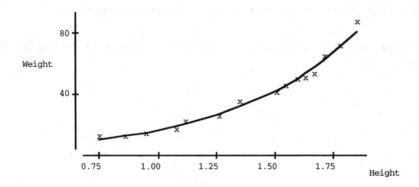

So there is some linearity in the data (as reflected in the 88% r^2), but our eyes tell us a better fit might be curved! How do we tell what the curve is? There are many approaches. The one we'll use here involves logs.

▧ Straightening out a Curve

If we suspect that the data might best be fit with a curve, we can try to straighten out the data and then fit a line. What could straighten out a data set? One common curve to try is an exponential curve like 2^x or 10^x.   Remember that logarithms have the really handy property that when applied to both sides of an equation like:

$$a = b^n$$

we get

$$\log(a) = n \log (b).$$

In other words, taking the log of a power makes it linear. So, to straighten out a power use a log. Here's what we try:

If the data (x, y) look curved, try graphing (x, log (y)).
If that looks straight, fit a line to it.

Let's try this with our data. We've been using weight as our y-variable, so let's build a new column containing log(weight):

ID	Height(m)	log(Weight)
1	0.75	1.07918
2	0.86	1.07918
3	0.95	1.14613
4	1.08	1.23045
5	1.12	1.34242
6	1.26	1.39794
7	1.35	1.54407
8	1.51	1.61278
9	1.55	1.66276
10	1.60	1.69897
11	1.63	1.70757
12	1.67	1.73239
13	1.71	1.81291
14	1.78	1.85733
15	1.85	1.94448

Note that this column was built using the log button on a calculator, which means we've used base 10. That will be important in a moment. Here is the height versus log(weight) graph:

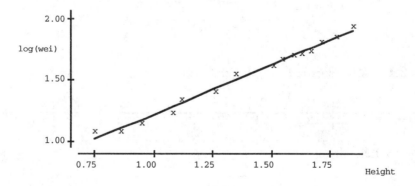

Hum! Much more "linear," and the $r^2 = 98.7\%$ (found using the computer). A very strong relationship; in fact, we could use this to predict! The "line" from this is:

$$\log(weight) = 0.401 + 0.815 \times Height$$

It would be nice to have this relationship changed back to one with weight instead of log (weight). In other words, we need to "unlog" the weight: How do we do this? Remember the definition of log base 10:

$$\log(x) \text{ is the power of 10 that is equal to } x$$
$$\text{or}$$
$$y = \log(x) \text{ if and only if } x = 10^y.$$

We'll use this definition to "unlog." We have something that looks like log y = x so we can "unlog" by replacing x with 10^x. Let's try this with our example:

$$\log(\text{weight}) = 0.401 + 0.815 \text{ Height}$$

$$\text{weight} = 10 \char94 (0.401 + 0.815 \text{ Height})$$

This is gives us a cumbersome exponent. We can simplify this a little more if we use the rule about exponents that says if we multiply like bases we add the exponents:

$$x^a x^b = x^{(a+b)}$$

We'll use this backwards on the 10 ^(0.401 + 0.815 Height) part:

$$10 \char94 (0.401 + 0.815 \text{ Height}) = 10 \char94 0.401 \times 10 \char94 (0.815 \text{ Height})$$
giving:
$$\text{weight} = 2.518 \times 10^{0.815 \text{ height}}.$$

Let's try to predict some weights given some heights. What is the predicted weight of a male with height 1 meter?

$$\text{weight} = 2.518 \times 10^{0.815 \,(1)} = 16.45 \text{ kg.}$$

Does this seem to fit? Well from the original data we know that a person who was 1.08 meters tall weighed 17 kg, so 16.45 seems reasonable.

What is the predicted weight for a male 1.95 meters tall? (That's pretty tall!)

$$\text{weight} = 2.518 \times 10^{0.815 \,(1.95)} = 97.79 \text{ kg.}$$

Example 5.1: Consider the following data set:

x	2	3	3	5	6	8	8	9
y	5	8	10	21	33	84	87	133

Fit a curve to these data.

Solution: Start with a graph, and notice that again we have a non-linear relationship.

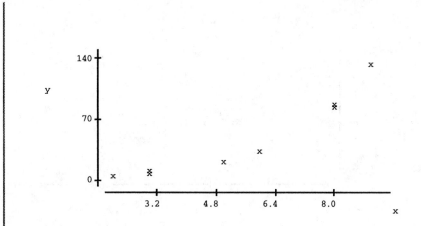

Let's try fitting a line to the logged data as before:

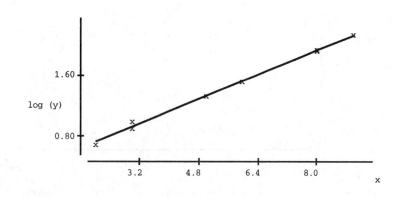

Nice! What was the regression line? log (y) = 0.330 + 0.200 x. Now we'll unlog it like before:

$$\log (y) = 0.330 + 0.200\ x$$
$$y = 10^{\wedge}(0.330 + 0.200\ x)$$
$$y = 10^{\wedge}0.330 \times 10^{\wedge}(0.200\ x)$$
$$y = 2.14 \times 10^{.2x}$$

Hum!

This technique doesn't always work. Consider the data set on the growth of a fly population in a closed chamber over time in the example below.

Example 5.2: Investigate the growth of the fly population over time given the data below. Can we predict the fly population at t = 12 or later?

time	1	2	3	4	5	6	7	8	9	10	11
flies	846	901	937	961	976	985	991	995	997	998	999

Solution: Let's start with a graph:

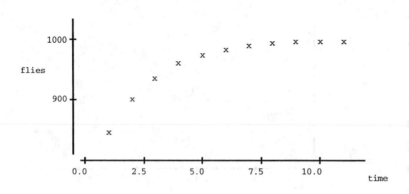

Definitely curved! Here's the graph of time and log (flies):

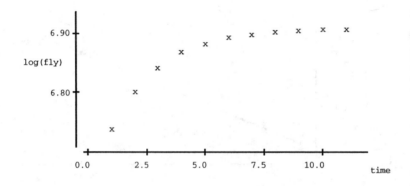

Still curved! So we would need to do something else to straighten this one out. Another common trick to try if just logging the y-variable doesn't work, is to try logging x and y both. Here is the graph of log(time) versus log(flies):

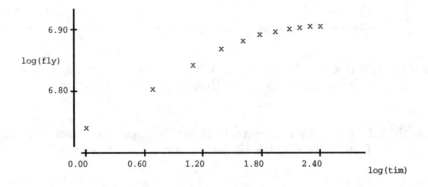

Not straight yet. So relying completely on this trick doesn't help. Can we predict future populations in this example? Look at the trend in the original graph; the population appears to be leveling off – perhaps the population has almost filled the container, and so fewer offspring live.

From the trend it seems safe to predict that the population is leveling off at around 1000.

Investigating Growth Over Time

Consider the population of the United States and the world given below. What would we predict the year 2000 populations to be?

Year	U.S. Population (Millions)	Year	world pop. (billions)
		1950	2.6
1960	179.3	1960	3.1
1970	203.3	1970	3.7
1980	226.5	1980	4.5
1990	250	1990	5.4

First consider the US population: Notice that the graph is almost exactly a straight line. Notice that the years are evenly spaced; we have a data point for every tenth year. This allows us another method for deciding if the graph is really a straight line.

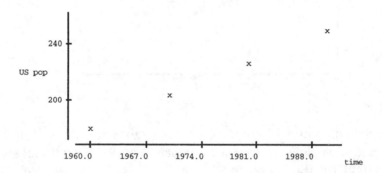

Look at the differences: from 1960 to 1970 24 million, from 1970 to 1980 23.3 million and from 1980 to 1990 23.5 million. Every ten years the population has increased by about the same amount, roughly 23.5 million. For even more evidence, when we do a regression using years as the independent variable and predicting population, we get an r^2 of .9999556. That means that a straight line is a very good model for these data. The equation gotten from the regression output gives the equation:

$$\text{population} = 2.353 * \text{year} - 4432.4$$

The amount that these data deviate from the line is negligible, probably less than the error in the population measurements. For instance the population in 1990 is listed as 250 million. 1990 times 2.353 plus -4432.4 is 250.07, so the line is very close to the actual population. We could use this line to predict the population in the future, say for the year 2000. We get

$$2000 * 2.353 - 4432.4 = 273.6$$

So in 2000 we predict 273.6 million as the U.S. population. You could predict far into the future too, but that would be stupid. Notice that the model says that population grows about 2.35 million each year.

We used the graph, the differences, and the regression output to decide whether a straight line is an appropriate model for these data. The increases should be more or less constant, the line should fit the graph well, with no pattern to the deviation from the line, and the r^2 should be close to 1. The idea with this model is that population is increasing by a fixed amount each year. Perhaps the basic population isn't growing at all, but there is a fixed amount of immigration each year or every ten years.

Now consider the World population, reprinted here with the increases:

Year	world pop. (billions)	increase
1950	2.6	
1960	3.1	0.5
1970	3.7	0.6
1980	4.5	0.8
1990	5.4	0.9

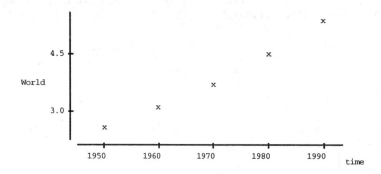

These data present a different picture. A straight line doesn't miss the points much, and if we do a regression we see an r^2 of .9855, so it looks as though a straight line might work. On the other hand the differences from one ten-year period to the one before are changing. They go from 0.5 to 0.9 in the four time periods. Another problem arises when we fit a line to these data. It will be below the first point, close to the second, above the third, close to the fourth and below the fifth. The line is close, but there is a pattern to the deviation from the line. It looks as though these data do not go up linearly, that is, at a constant rate, but increase at an increasing rate. There must be more happening. Let's compute the ratios of one time to the one before.

Year	World Pop (billions)	increase	new/old
1950	2.6		
1960	3.1	0.5	1.1923
1970	3.7	0.6	1.1935
1980	4.5	0.8	1.2162
1990	5.4	0.9	1.2

For example, the new/old ratio for 1990 is gotten by taking the ratio of 5.4 to 4.5, giving 1.2. That is like a 20% rate of growth for the entire ten-year period. This looks better than the model that assumes a constant amount of growth each year. This model assumes that the amount of growth is proportional to the population. This is a very popular model for population growth, since it seems reasonable to assume that the number of deaths and the number of births are both proportional to the population and the growth equals births - deaths if there isn't any immigration.

From this we see that the differences grow, but the ratios are almost the same. So we can go from one period to the next by multiplying by that ratio. There are more sophisticated ways to obtain a good choice of a ratio, but we will just estimate. 1.2 seems like a good choice. That means that we are assuming that the population grows by a constant percentage, 20% each 10 years. We therefore estimate the population in the world in the year 2000 to be 5.4*1.2 = 6.46 billion. For 2010 we would estimate 6.46*1.2 = 7.776 billion. We could go directly from 1990 to 2010 by multiplying 5.4 times $(1 + .2)^2$ as we did with compound interest.

We could call this a 20% growth rate for each ten-year period, but we would like an annual rate. This could be phrased back into our compound interest language as - what annual interest rate would give a 20% increase in an investment after 10 years? Solving the equation $(1 + r)^{10} = 1 + .20$ by using trial and error or via logs we get r = .018. So the world's population is growing at a rate of almost 2% per year.

We can use this same approach to estimate the annual interest rate if we have some data on total amount in the account

Example 5.3: Here's some information about the amount of money in an account over a period of 8 years. Can we estimate the interest rate?

year	1	2	3	4	5	6	7	8
$$	1080	1166	1260	1360	1469	1586	1714	1851

Solution: Clearly these data are not linear. Let's try the approach using with the world population. We need to find the new/old ratios:

$$1166/1080 = 1.0796$$
$$1260/1166 = 1.0806$$
$$1360/1260 = 1.0794$$
$$1469/1360 = 1.0801$$
$$1586/1469 = 1.0796$$
$$1714/1586 = 1.0807$$
$$1851/1714 = 1.0799$$

Looks like 1.08 or there abouts is the ratio, so we are looking at using a curve with equation

$$\$\$ = P (1 + 0.8)^{year}.$$

Can we find P?? Sure! We know that after 1 year we have 1080, so
$$1080 = P (1 + 0.8)^1$$
$$P = 1000$$

YOU TRY IT 5.1

Investigate these data on the population of Kenya. Is the population growth exponential like the world population? Find the annual growth rate.

year	population
1955	7179
1960	8332
1965	8749
1970	11498
1975	13741
1980	16632
1985	20353
1990	25130
1995	30844

One Last Word

There is no one method for dealing with non-linear data. We have examined two tricks that work well for data that are growing exponentially like money does and some populations do. If you must deal with non-linear data, use your common sense! Don't resort to fitting a line if the data are clearly non-linear.

Exercise Set 3.5

1) Investigate the change in the population (in millions) of Florida over time. Is it linear? Exponential? What is the growth rate? What does this predict for 1990 (actual population = 12.94 million)?

year	1900	1910	1920	1930	1940	1950	1960	1970	1980
pop	0.529	0.753	0.968	1.468	1.897	2.771	4.952	6.791	9.747

2) Investigate the change in the population (in millions) of Alabama over time. Is it linear? Exponential? What is the growth rate? What does this predict for 1990 (actual population = 4.041 million)?

year	1900	1910	1920	1930	1940	1950	1960	1970	1980
pop	1.829	2.138	2.348	2.646	2.833	3.062	3.267	3.444	3.894

3) Investigate the change in the population (in millions) of California over time. Is it linear? Exponential? What is the growth rate? What does this predict for 1990 (actual population = 29.76 million)?

year	1900	1910	1920	1930	1940	1950	1960	1970	1980
pop	1.485	2.378	3.427	5.677	6.907	10.59	15.72	19.97	23.67

4) Investigate the relationship between the distance of a planet from the sun and the period (time for one rotation around the sun).

planet	period days	distance millions of miles
Mercury	88	36.0
Venus	225	67.2
Earth	365	92.9
Mars	687	141.5
Jupiter	4329	483.3
Saturn	10,753	886.2
Uranus	30,660	1782.3
Neptune	60,150	2792.6
Pluto	90,670	3668.2

5) Estimate the annual interest rate from the data below:

year	1	2	3	4	5	6
$$	1240	1546	1950	2410	3050	3710

6) Estimate the annual interest rate from the data below:

year	1	2	3	4	5	6
$$	10,042	10,100	10,133	10,170	10,240	10,260

7) Investigate the relationship between temperature °F and chirping for the two different cricket species given below.

striped ground crickets:

Temp	89	72	93	84	81	75	70	82	69	83	80	83	81	84
Chirps	78	60	79	73	68	62	59	68	61	65	60	69	64	68

snow tree crickets:

temp	50	55	60	65	70	75	80
chirps	40	60	80	100	120	140	160

CHAPTER SUMMARY   6

Now that we have practiced some basic techniques of data analysis, we need to learn to recognize when to apply which technique. Rarely will you encounter data along with the directions: *"Generate a box plot and use it to discuss..."* It is much more likely that you will encounter a question about a population for which you'll have to decide how to collect the data, which kind of graph best illustrates the data, and what analysis tool would be most appropriate for answering the question. Deciding which way to proceed will most likely involve trial and error, involving several styles of graphs or analysis techniques.

Comparison or Relationship?

One of the first decisions that you will make after collecting the data for your analysis involves deciding whether you are looking at a single numerical characteristic for the population or are you investigating a relationship between two (or more) characteristics? Depending upon the data collected, you may need to investigate both of these things.

On the next page is a data set from 23 students who took this class in 1997. Two of the columns concern short term memory. The column REAL is the number of three letter words each student recalled after examining a list of 20 three letter words. The words were covered after 30 seconds, and the students wrote down as many as they could remember. The experiment was repeated with 30 nonsense triples, like zoq. The number of nonsense triples recalled is stored in the column NONS.

Suppose we want to investigate the data on real words remembered, nonsense words remembered and gender to see if there are memory differences between the sexes and whether there's a relationship between remembering real words and recalling nonsense triples. How should we start?

ID	SEX	SMOKE	HEIGHT	ARM	FOOT	HEAD	SIBS	CLASS	DIST	TIM	REAL	MONS	BREA
1	H	Y	188	194	31	59.5	1	FR	200	65	12	7	58
2	H	Y	189	179	30	56	3	SO	180	57	7	3	50
3	H	Y	186	185.5	28	60	0	SO	40	61	9	5	51
4	H	N	172	175	28	57	2	FR	60	51	5	2	59
5	H	N	177	175	29	59	0	SO	100	65	7	4	31
6	H	N	179.5	180	31	59	1	FR	110	68	4	3	35
7	H	N	190	188.5	32	61	4	FR	90	70	6	3	49
8	H	N	168	165	27	52	3	SO	90	56	9	2	62
9	F	Y	162	160	24	55	2	SO	120	55	7	6	35
10	F	Y	184	182	29	54.5	1	FR	250	60	5	4	15
11	F	Y	167	163	25	53	1	FR	310	42	6	3	32
12	F	Y	157.5	157.5	24	54	0	FR	90	42	7	3	55
13	F	N	169.5	165	24	53	1	SO	110	120	8	3	20
14	F	N	164	156	25	55.5	1	FR	90	46	11	6	35
15	F	N	174	163	25	53	2	FR	140	58	10	2	65
16	F	N	159	152	25	54.5	1	JR	30	25	6	4	30
17	F	N	164.5	173	26	52	2	FR	200	55	7	5	51
18	F	N	169	174	28.5	56	1	FR	90	61	7	4	40
19	F	N	166	165	25	55	1	FR	70	64	11	2	66
20	F	N	176.5	171	26.5	55	1	SO	210	35	9	3	30
21	F	N	163.5	158	26	50.5	1	SO	100	42	11	3	57
22	F	N	161	146	25	51	3	FR	90	62	7	3	30
23	F	N	162	163	24	50	5	FR	50	45	8	7	37

When looking at differences between genders, we are really interested in a single column of data -- such as REAL. We then want to subdivide the column into male versus female and compare. We used several tools to do this, most notably box plots and histograms. Remember the general rule:

> **Use box plots when investigating one variable**
> **or how two groups differ on one variable.**

The other issue in this example is to ask whether real words recalled is related to nonsense words recalled. So we are trying to look at each person's pair of data to see if there's a pattern. We relied on scatter plots to handle situations like this, where we were really looking at ordered pairs of data, like real and nonsense words, height and arm span, population and time, etc.

> **Use scatter plots and regression when investigating**
> **how pairs of variables are related.**

One Final Word

There is no one right way to summarize and analyze data, but there are many wrong ways! The ideas in this chapter will help to make you a better consumer of statistical information.

Key Ideas

Collecting Data

- Use random samples whenever possible.
- Try to make survey questions as un-biased as possible.

Presenting Data

- Choose a graph style that best illustrates the aspect of the data you are interested in.
- Avoid the common mistakes of bad scaling and misuse of area.
- Correct histograms have equal sized classes.

Summarizing Data

- Measures of Center: The mean is the average and the median is the middle value.
- Measures of Spread: The standard deviation goes with the mean and the quartile calculations go with the median.
- The lower quartile is the middle of the bottom half of the data and the upper quartile is the middle of the top half of the data – *after* the data have been sorted.
- Skew: We use a qualitative measure of skew. If the data are clumped towards the lower end and have a long tail at the higher end, we say the data are skewed right. If the opposite is true we say the data are skewed left.

Finding Relationships

- Scatter plots are drawn to check on relationships between two variables.

- The r^2 value is always between 0 and 1 and can be interpreted as a % goodness of fit. The closer to 1 the more linear the relationship:
  - strong: $65\% < r^2 \leq 100\%$,
  - moderate: $25\% < r^2 \leq 65\%$,
  - weak: $10\% < r^2 \leq 25\%$.

- Not all relationships are linear; we can sometimes find the curve for a non-linear relationship using logs or, if the data are evenly spaced, examining the ratio of consecutive entries.

Summary Exercises

Problems 1 through 6 refer to the data set on page 219. The questions are deliberately open ended. Don't be afraid to try different approaches.

1) Consider the data on gender and real and nonsense words remembered from the class data. Graph these data in ways that seem reasonable. Using your graphs as evidence, and also using any of our statistical calculations, draw appropriate conclusions on relationships you see in these data.

2) Consider the data on gender, smoking and breath-holding. Graph these data in ways that seem reasonable. Using your graphs as evidence, and also using any of our statistical calculations, draw appropriate conclusions on relationships you see in these data.

3) A researcher has asked if there is any relationship among smoking and height. Does gender play a role? Help her out using the class data, some graphs and some words.

4) It seems unlikely, but a researcher has speculated that class (freshman, sophomore, junior, senior) is related to distance from home. Can you produce some evidence one way or another from the class data and draw a conclusion?

5) Do men have more siblings than women? Do smokers have more siblings than non-smokers?

6) Someone has guessed that people who think time is going by quickly are likely to not hold their breath very long. Using the class data MIN and BREA try to confirm or deny this guess. Maybe if we also took smoking into account there would be more of a pattern.

7) Investigate the data below on the stock market, illustrating the data with a suitable graph and discussing any trends. Can you find a way to predict future values based upon these data?

Year	Dow
1980	839
1981	964
1982	875
1983	1047
1984	1259
1985	1212
1986	1547
1987	1896
1988	1939
1989	2169
1990	2753
1991	2634
1992	3169
1993	3301
1994	3758
1995	3834
1996	5177

4   LINEAR PROGRAMMING

Successful manufacturers are very concerned with **maximizing** their profits or **minimizing** their costs. Companies must plan carefully how many of their products to make and how much to sell them for. They have **constraints** on numbers of employees and equipment and storage space for or availability of raw materials to consider as well. This is not done by trial and error; a good company will attack the problem very systematically, often using an approach called **Linear Programming**. We'll find that linear programming problems can be small enough to do easily by hand or so large that specialized software packages on large computers are needed. In this chapter, we will learn about setting up and solving realistic linear programming problems.

After completing this chapter, you will be able to recognize linear programming problems and translate them into mathematics. You will be able to solve problems of any size on the computer and solve simple two-variable problems by hand. You will also be able to interpret the output from the computer program to determine what would happen if the situation presented in the problem should change.

The History of Linear Programming

Linear Programming is a relatively new approach. It grew out of troop supply problems that arose during the second world war. With a war on several fronts, the shear size of the problem of coordinating supplies was daunting. Mathematicians had mobilized for the war effort and teams of very bright people set to work on this problem. They looked for an approach that could make use of computers, which were being developed at that time and offered the possibility of doing many simple calculations quickly. Finally, in 1947 **George Danzig** developed the simplex algorithm for solving linear programming problems. The simplex algorithm is a simple recipe for solving linear programming problems of any size and is very easy to program on a computer. Since 1947 many other ways of solving linear programming problems have been developed, mostly for problems that have special forms such as those that contain two independent parts.

In recent years linear programming has been applied to problems in almost all areas of human life. A quick check of our library found entire books on linear programming applied to agriculture, city planning, business, and several other areas. Even feed lot operators decide how much of what kinds of food they should give their livestock each day with linear programming. Most computing facilities have software to solve even the largest linear programming problems.

What Makes a Problem Linear?

To solve a linear programming problem we will need to **recognize** the problem as "linear" and then **translate** the English description of the problem into a mathematical one. Mastering this skill takes practice. To illustrate what a typical linear programming problem looks like, let's consider the

Barbie and Ken Problem

A doll factory wants to plan how many Barbie and Ken dolls to manufacture in a week to maximize company profit. A Barbie doll earns $6.00 in profit and is made of 12 ounces of plastic for her body and 5 ounces of nylon for her hair. A Ken doll earns $6.50 and is made of 14 ounces of plastic. Each doll goes in a box made of 4 ounces of cardboard. The company can only get one weekly shipment of raw materials, including 100,000 ounces of plastic, 30,000 ounces of nylon and 35,000 ounces of cardboard. The marketing department has researched the sales potential of the dolls and found that children buy at least 2 Barbies for each Ken.

What makes this a linear programming problem? On the surface this looks like a simple word problem from algebra. Look carefully at the words and notice some differences:

1) It looks like there are many workable choices for how many Kens and Barbies to make, anywhere from none at all to thousands.

2) Somehow the weekly shipment amounts of plastic, cardboard and nylon restrict what we can do, as does that sentence about twice as many Barbies as Kens.

3) Instead of trying to find just one solution we are asked to find the "best" solution, where in this case best means the one that makes the company the most money.

For a problem to be a linear programming problem it must have the properties listed below. Each situation must be examined to see if assuming these ideas is reasonable. Occasionally the assumptions aren't completely met, but are reasonable approximations to reality, so we use the linear programming approach anyway.

Proportionality In the Barbie and Ken problem it makes sense that twice as much plastic makes twice as many dolls. That's what proportionality means, that it takes proportionally more material to make proportionally more of the product. Profits must also increase in the same proportion as the number of items sold increases. This would be violated if we gave Walmart a quantity discount.

Divisibility We must be able to produce fractional parts of items. In the Barbie and Ken problem this appears to be violated, because we can't sell two-thirds of a Barbie doll. On the other hand we can interpret making 2/3 of a doll as making two dolls over a three week period, so we can work around this one.

Short Time Linear programming is a short term tool. Labor constraints aren't forever; we can hire and train new people. Sales constraints can be changed by advertising campaigns. Costs

change over time. In the Barbie and Ken problem we are
dealing with weekly production figures, which is OK.

These properties are easier to verify than they first appear; linear
programming problems will be easy to recognize. Now let's see how to pull the
mathematics from the English.

1   FORMULATING THE PROBLEM
– Translating from English to Math–

Let's learn about translating by working through the Barbie and Ken
example first and then building a general recipe for translating. The first task is to
decide what quantities we are asked to find. The Barbie and Ken problem asks us
that in the first sentence - "How many Barbie and Ken dolls ...?" To find these
quantities is to find the solution, so we let B represent the number of Barbies to
make and K represent the number of Kens to make. Then we set up the rest of the
problem using B and K as the unknowns. We state this mathematically as:

```
B = the number of Barbies to make per week
K = the number of Kens to make per week
```

Next we need to decide how to use the information in the other sentences.
The **objective** of a linear programming problem will always be to maximize or
minimize a quantity. In the first sentence of the problem we see "maximize
company profit." Profit is $6.00 for each Barbie and $6.50 for each Ken, so the
total amount of money, Z, that they get will be $6B + 6.50K$. We can state this as:

```
Maximize Z = 6B + 6.5K
```

Now we have to translate all the limiting conditions or **constraints**. It helps
to make a table summarizing the numbers associated with the constraints as an
intermediate step. Include a row for each constraint and a column for each
unknown, with one more column for the limits on the constraints. For example,
12 is the number associated with plastic and Barbie so it goes in the plastic row
under B. We have left the demand row blank for now.

	B	K	limit
plastic	12	14	100,000
cardboard	4	4	35,000
nylon	5		30,000
demand			

1)   **Plastic** is in short supply; the information about plastic says we must use
less than or equal to 100,000 oz. The chart gives 12 oz per B and 14 oz per K.
Thus **12B** is the amount of plastic we use in making Barbies. Similarly, **14K** is
the amount of plastic we use in making Kens. So 12B + 14K is the total amount
of plastic used, and it must be less than 100,000. We can state this as:

```
12B + 14K ≤ 100,000
```

2)   **Cardboard** is in short supply just like the plastic. Let's proceed the same
way. The chart gives 4 oz of cardboard each, so we need 4K for the Kens and 4B
for the Barbies. There are **at most** 35,000 ounces, so:

```
4B + 4K ≤ 35,000
```

3)   **Nylon** is used only in Barbie dolls, at 5 oz. per doll. That gives:

$$5B \leq 30,000$$

4)   There is one last **demand** restriction to translate. We sell **at least** two Barbies for each Ken.   10 Barbies and 5 Kens are fine. 20 Barbies and 5 Kens are too. 10 Barbies and 6 Kens won't work. Try a few numbers of your own:

B	K	Inequality
10	5	True
20	5	True
10	6	False
___	___	___
___	___	___
___	___	___

Sometimes it is easy to spot the pattern from a table like this: If we multiply the number of Kens by 2 there must be at least that many Barbies. In symbols that is $B \geq 2K$. We will rearrange it slightly so that the variables are both on the left side:

$$B - 2K \geq 0$$

These kinds of sentences can be difficult to translate. An alternate approach is to think in terms of proportions. Look at the wording in the problem again:

"...children buy at least 2 Barbies for each Ken."

Another way to say the same thing more mathematically is:

"The ratio of Barbies to Kens is at least 2 to 1."

Remembering that a ratio is a fraction, and "at least" means greater than or equal to, we get:

$$\frac{B}{K} \geq \frac{2}{1}.$$

Cross-multiplying:

$$B \geq 2k, \text{ or } B - 2k \geq 0$$

Which is the same as we got by guessing the pattern from the table.

We now fill in the last row of the table to give:

	B	**K**	**limit**
plastic	12	14	100,000
cardboard	4	4	35,000
nylon	5	0	30,000
demand	1	-2	0

Notice that each constraint is an inequality rather than an equality. We have a certain amount of plastic, cardboard and nylon; we can't use more than we have, but using less is OK. The complete mathematical translation is below. The last pair of constraints is to remind us that making negative numbers of dolls is not realistic. Even the obvious must be stated in the mathematical formulation.

```
        B = the number of Barbies to make per week,
        K = the number of Kens to make per week.
Maximize Z =      6B + 6.5K
Subject to:      12B +  14K ≤ 100,000   (plastic)
                  4B +   4K ≤ 35,000    (cardboard)
                  5B         ≤ 30,000    (nylon)
                   B -   2K ≥ 0         (sales)
                   B ≥ 0 and K ≥ 0      (non-negativity)
```

Later we will solve this problem using algebra; for now we'll experiment with different values for B and K, and check to see if they satisfy the constraints.

Let's start with the obvious. We can make no Barbies and no Kens. That would satisfy all the inequalities, and it would make no profit. The **point** would be (0,0), where the first coordinate is the number of Barbies to make and the second coordinate is the number of Kens to make.

A point that satisfies all the constraints is called **feasible**. The **feasible region** is the set of points that satisfy all the inequalities, in other words all the points that are feasible. We already found that (0,0) is feasible. Below are two more points that are checked against the constraints for feasibility.

Point	Plastic	Cardboard	Nylon	Demand
(B, K)	$12B + 14K \leq 100,000$	$4B + 4K \leq 35,000$	$5B \leq 30,000$	$B - 2K \geq 0$
(1000,1000)	True	True	True	False
(5000,2000)	True	True	True	True

From this table we see that 1000 Barbies and 1000 Kens is not a possible answer to our problem, but 5000 Barbies and 2000 Kens could be, since all of the constraints are true. *Is* it the answer? Let's look more closely at the constraints with B = 5000 and K = 2000:

$$12B + 14K = 88,000 < 100,000$$
$$4B + 4K = 28,000 < 35,000$$
$$5B = 25,000 < 30,000$$
$$B - 2k = 1000 > 0$$

Notice that all of these inequalities have room left in them, i.e., after making 5000 Barbies and 2000 Kens, we still have plastic, nylon, and cardboard left over, and the demand also has room. Common sense says that we ought to be able to make more dolls, provided we don't go over the limit for one of the constraints! In fact, if only one of the constraints has been pushed to the limit, we might even be able to make careful changes to B and K to keep the first one at the limit and make another constraint go to the limit. This leads us to the idea:

**We won't have a solution until we run out of room in
some of the constraints.**

This is the fundamental idea behind the procedure for solving these problems. Now it's your turn to check some possible feasible points.

YOU TRY IT 1.1

Test 5 more points for feasibility in the Barbie and Ken Problem. See if you can find points that satisfy all of the constraints, and use up all of the room in at least one constraint. Compute the profit for each feasible point as well.

(B, K)	$12B + 14K$ $\leq 100,000$	$4B + 4K$ $\leq 35,000$	$5B \leq 30,000$	$B - 2k \geq 0$	Profit $6B + 6.50K$

If we could build a table like the one above, listing **all** the feasible points that "use up" at least two of the constraints, then we could solve the problem by choosing the point with the biggest profit. Unfortunately, since we're allowing fractional numbers, there are infinitely many feasible points. We'll need to find another approach to solve these problems. Before learning more about solving, let's practice translating.

Translating a Linear Programming Problem
1. **Read** the problem carefully. Underline key words, then read it again.
2. Identify the **variables** and write down what they stand for. The Barbie and Ken problem asked " How many Barbies and Kens should they manufacture each week to maximize their profit?" That meant that the number of Barbies and the number of Kens were the variables. Remember that variables are things that can change, so if you have a **fixed** amount of something it can't be a variable. Name variables with letters that signify what they stand for, like B for the number of Barbies. Write out the units for the variables.
3. Find the **objective function**. Normally, there is a cost or profit associated with each unknown and the objective function is composed of sum of the value per item times the variable for that item, like 6.5K is the profit per Ken times the number of Kens. Is it a maximization or a minimization?
4. Find the **constraints**, things that restrict, or limit. If a process uses hours of labor, time on a machine, material, or anything else that is in short supply, it will generate a constraint. Summarize the constraint quantities in a table, giving each constraint a name. If there are "300 available" expect a constraint: something ≤ 300. If there must be "at least 200" expect: something ≥ 200. The left-hand-side will be a sum of numbers times variables, where the numbers are how much of the limited item it takes to produce one unit of the variable.
5. Go over the model and the words. **Be careful about units**. Each inequality must have consistent units between the terms on the left-hand-side and the right-hand-side. If hours and minutes both appear in the problem decide which to use and change everything to that. Carefully consider each constraint in words and see that it is expressed correctly as an inequality.

Example 1.1: Let's translate the following problem, following the steps.

The manufacturing process at a particular oil refinery requires it to produce at least 2 gallons of gasoline for each gallon of fuel oil. To meet the winter demand for fuel oil, at least 3,000,000 gallons a day must be produced. The demand for gasoline is no more than 8 million gallons a day. If the refinery makes $0.88 for each gallon of gas and $0.65 for each gallon of fuel oil, how much of each should be made to maximize profit?

Solution: Here we go!

1) Read the problem and underline key words such as "at least" and "maximize". Then carefully read it again.

2) "How much of each" implies that the unknowns are amount of gas (G) and the amount of fuel oil (F) manufactured per day in gallons.

3) "Maximize profit" implies that we want to maximize

0.88G + 0.65F.

4) There are 3 constraints: oil demand, gas demand, and production.

Oil demand: "at least 3,000,000 gallons a day" implies that there is a fuel oil demand constraint. In other words: amount of fuel oil is at least 3,000,000 or **F ≥ 3,000,000.**

Gas demand: "no more than 8 million gallons per day" implies that there is a gasoline demand constraint. In other words: amount of gasoline is at most 8,000,000, or **G ≤ 8,000,000.**

Production: "at least 2 gallons of gasoline for each gallon of fuel oil" implies that we have a processing constraint. This can be phrased as a proportion: "the ratio of gas to oil is at least 2 to 1," so

$$\frac{G}{F} \geq \frac{2}{1}.$$

Cross multiplying gives G ≥ 2F, or **2F - G ≤ 0.** Now we can fill in the table:

	F	G	limit
production	2	−1	0
oil demand	1	0	3,000,000
gas demand	0	1	8,000,000

5)   Now we're ready to state the mathematical formulation.

```
G = number of gallons of gas made in one day
F = number of gallons of fuel oil made in one day

        Maximize          Z = 0.65F + 0.88G
        Subject to:
          production:      2F -  G ≤ 0
          oil demand:       F      ≥ 3,000,000
          gas demand:            G ≤ 8,000,000
                      F ≥ 0, G ≥ 0
```

Here are two problems for you to try your new-found skills on. Be careful; the second one is more complicated.

YOU TRY IT 1.2

Translate the following problem, following steps 1 through 5:

A dietitian wishes to prepare a meal which has a minimum of calories but still satisfies nutritional requirements. In particular it must have at least 3 oz of protein, 1 oz of fat, and 3 oz of carbohydrates. The foods available are: a salad which has 30 cal/oz and is 20% protein, 20% fat, and 60% carbohydrates; and meat which has 70 cal/oz, is 50% protein, 30% fat, and 20% carbohydrates. What amounts of salad and meat should she serve?

1) Read carefully and underline. Read again.

2) What are the variables? What are the units on the variables?

3) What is the objective? Are we maximizing or minimizing?

4) What are the constraints?

5) Give the complete mathematical formulation, including all inequalities.

YOU TRY IT 1.3

Translate the following problem, following steps 1 through 5:

As part of a campaign to promote its Annual Sale, the Excelsior Company decided to buy television advertising time on Station KAOS. Excelsior's television advertising budget is $102,000. Morning time costs $3000 per minute, afternoon time costs $1000 per minute, and evening time costs $12,000 per minute. KAOS cannot offer Excelsior more than 6 minutes of evening time or more than a total of 25 minutes of total advertising time over the two weeks in which the commercials are to be run. KAOS estimated that each minute of morning commercials would be seen by 200,000 people, afternoon commercials by 100,000 people, and evening commercials by 600,000 people. How much morning, afternoon, and evening time should Excelsior buy to maximize exposure of its commercials?

Exercise Set 4.1

In problems 1 through 4 below, given the variable names translate each part into one or more constraint(s) in those variables.

1) Let P = number of pound of peanuts to be put in a nut mixture and A be the number of pounds of almonds to be put in a nut mixture. Translate:

a) The nut company has at most 800 pounds of peanuts and 600 pounds of almonds available each day for the mixture.

b) The company wants at least three times as many almonds as peanuts in their mixture.

c) A pound of peanuts must be dry roasted for 4 hours prior to mixing and a pound of almonds must be dry roasted for 2 hours prior to mixing. The company has at most 200 pound hours of dry roasting each day.

2) Let S = the number of gallons of shellac to produce and V = number of gallons of varnish to produce. Translate:

a) One gallon of shellac requires 2 ounces of organic gum and one gallon of varnish requires 5 ounces of organic gum. The total amount of organic gum available is 10 gallons.

b) Consumer demand limits the number of gallons for varnish to at most half the number of gallons of shellac.

3) Let WW = gallons of white wash, DC = gallons of deck cleaner and DS = gallons of driveway sealer. Translate:

a) We have at most $500.00 for refurbishing our cabin and white wash costs $16 per gallon, deck cleaner $8 and driveway sealer $12.

b) We have room in the pickup to carry at most 25 gallon cans.

c) We have at most 200 square feet of driveway to seal and at least 1000 square feet of deck to clean. One gallon of driveway sealer seals 50 square feet, and one gallon of deck cleaner cleans 150 square feet.

4) Let G = goose down jackets, W = wool jackets, C = cotton jackets and T = thinsulate jackets. Translate:

a) We have at most 1200 pounds of wool, 1500 pounds of cotton, 1400 pounds of down and 1600 pounds of thinsulate filling. Each jacket requires 60 ounces of fill.

b) Market demand studies indicate that we need at least twice as many thinsulate jackets as all others combined.

Translate each of problems 5 through 23. Be sure to write descriptions of the variables in English and label all the inequalities.

5) A chocolate bar contains 50 calories and 5 grams of protein, and makes a child smile 7 times while eating it. A granola bar contains 25 calories and 15 grams of protein, and makes a child smile 2 times while eating it. A teacher plans to serve some of these snacks to his first grade class. They should total at least 40 grams of protein and no more than 150 calories. How many of each kind of bar should he serve to maximize smiles?

6) Sarah has decided to try to make some money by making and selling mailboxes. She has contacted a store that will buy all she can produce. She makes regular mailboxes and deluxe mailboxes, and makes a profit of $9 on the regular mailboxes and $16 on the deluxe mailboxes. It takes an hour to make a regular mailbox and 3 hours to make a deluxe mailbox. Her work area is so small that she can only paint 7 total mailboxes a day. The store wants at least 3 mailboxes each day. Sarah works hard, but a 10 hour workday is about all she can manage. How many of each type of mailbox should she make to maximize profit?

7) Mary wants to sell fruit punch and lemonade on the sidewalk in front of her house. In conference with her mom, she has made the following decisions:

Selling price: Fruit punch at 15 cents per cup

Lemonade at 12 cents per cup

In addition, Mary has estimated that her cost to provide a cup of fruit punch is 5 cents per cup and to provide a cup of lemonade is 3 cents per cup. Mary has a total of 40 cups for her operation, and available funds of $1.50. How many of each type of drink should she sell to maximize profit?

8) A manufacturer makes 2 kinds of cars, the sedan model and the sports model. According to a contract, it must supply at least 5000 cars. The sports model has a special engine of which only 2500 are available. Both models are made on an assembly line on which only 15,000 hours of time are available. The sedan takes 2 hours on the line and the sports car 4 hours. The profit is $750 on the sedan and $2000 on the sports model. How many of each should be made to maximize profit?

9) The Boone Weavers make shawls and afghans. They spin yarn, dye yarn and weave yarn for each. A shawl requires 1 hour of spinning, 1 hour of dyeing and 1 hour of weaving. An afghan requires 2 hours of spinning, 1 hour of dyeing and 4 hours of weaving. In a week, there is time to spend at most 13 hours spinning, 10 hours of dyeing and 30 hours weaving. How many shawls and afghans should be made to maximize profit if shawls bring in a profit of $25 and afghans a profit of $40?

10) Joe's Bottlers sells cola and uncola drinks in vending machines. Each machine will hold at most 150 bottles. A bottle of cola brings in $0.25 profit, and a bottle of uncola $0.45. Since more colas than uncolas are sold, they want at least twice as many colas as uncolas, but there should be at least 20 of each kind in a machine. How many of each should be in a machine to maximize profit?

11) An office manager needs to purchase storage lockers. She knows that a regular locker costs $40, requires 4 square feet of floor space, and holds 10 cubic feet of materials. On the other hand, a large storage locker costs $100, requires 8 square feet of floor space, and holds 20 cubic feet. Her budget permits her to spend at most $600, while the office has room for no more than 52 square feet of lockers. The manager desires the greatest cubic feet of storage capacity within the limitations imposed by funds and space. How many of each type of locker should she buy?

12) The State Employee's Credit Union (SECU) has $15.5 million in funds

available for the mortgages and auto loans applied for by it s members. The primary stockholders have recently approved the following policies regarding lending:

- Keep at least 5 times as much money invested in mortgages as cars
- Keep 18% of existing capital in short-term investments (i.e., not in loans).
- For the next quarter, set interest rates at 10.25% for car loans and 8.25% for mortgages.

Current loan applications include $4.2 million in car loans and $10.5 million in mortgages. How much should be allocated to each type of loan for the SECU to maximize its earnings?

13) A nut company sells three mixes, Regular, Deluxe and Supreme, each with a $4 profit. Regular has 50% peanuts, 30% cashews and 20% hazelnuts. Deluxe has 30% peanuts, 40% cashews and 30% hazelnuts. Supreme has 20% peanuts, 40% cashews and 40% hazelnuts. If the company has at most 800 pounds of peanuts, 400 pounds of cashews and 295 pounds of hazelnuts available in a day, how many pounds of each mix should it make?

14) A textile plant processes 2 types of cloth, an acrylic-cotton (AC) blend and a nylon-cotton (NC) blend. Each is blended 50-50. Acrylic and nylon are available in unlimited supply, but due to a recent drought, only 60000 pounds of cotton are available each month. The spinning department requires 1 hour of production time per 1000 pounds of AC blend and 2 hours per 1000 pounds of NC blend. This department has 150 available hours per month. The dye room has 150 available hours per month. The AC blend requires 0.5 hours per 1000 pounds in the dye room and the NC blend requires 3 hours. If the profit from the AC blend is $40 per 1000 pounds and $50 per 1000 pounds of the NC blend, how much of each blend should be produced in a month to maximize profits?

15) The Coswell Coat Company (CCC) makes 3 kinds of jackets: Cotton-filled (CF), wool-filled (WF) and goose-down filled (GF). Each involves 50 minutes of sewing, but it takes 8 minutes to stuff a CF, 9 minutes to stuff a WF and 6 minutes to stuff a GF. The factory has a supplier that will furnish it 225 pounds of wool and 135 pounds of goose-down and all the cotton that is needed per week. It takes 12 ounces total to fill a jacket. The factory has 666 and 2/3 hours of sewing time available and 80 hours of filling time available per week. It makes $24 on each CF jacket, $32 on each WF and $36 on each GF. How many of each should be made to maximize profit?

16) A manufacturing firm wants to consider making 3 new styles of recliners: the deluxe, the easyrider and the economy model. The times left over from other products in the various shops are given:

shop	available time (hours per week)
woodwork	500
finishing	500
upholstery	300

The number of machine hours needed to make one of each type of recliner are:

shop	deluxe	easyrider	economy
woodwork	10	4	5
finishing	4	5	0
upholstery	1	0	3

The sales department did a market survey and found that the sales potential for the deluxe and easyrider is more than the company can make in a week, but the maximum sales potential for the economy is 25 per week. The profit for each is:

chair:	deluxe	easyrider	economy
profit:	$60	$50	$25

How many of each recliner should the company make in order to maximize profit?

17) Mary, who is ill, takes mineral supplements. Each day she must have at least 21 units of iron, 10 units of potassium, and 40 units of calcium. She can choose between pill #1 which contains 7 units of iron, 2 of potassium, and 4 of calcium, and pill #2 which contains 3 units of iron, 2 of potassium, and 14 of calcium. Pill #1 costs $0.20 and pill #2 costs $0.35. Find the number of each pill which would minimize cost for Mary.

18) A dietary manager at a retirement center plans to serve a lunch of roast beef and bread. She is required by law to serve at least 400 calories and 3 milligrams of iron, but would like to minimize her cost. She can buy roast beef for 8¢ per ounce (100 calories and .8 milligrams of iron) and bread for 1¢ per slice (25 calories and .1 milligram iron). Find the amounts of roast beef and bread she should serve.

19) A small generator burns two types of fuel, low sulfur and high sulfur, to produce electricity. In one hour of use each kilogram of low sulfur fuel emits 3 units of sulfur dioxide, generates 3 kilowatts and costs 75 cents, while each kilogram of high sulfur fuel emits 6 units of sulfur dioxide, generates 3 kilowatts and costs 50 cents. The EPA insists that the maximum amount of sulfur dioxide emitted per hour be 9 units. At least 12 kilowatts must be generated per hour. How many kilograms of each fuel should be used hourly to minimize the cost of the fuel used?

20) Danbury Mining Corporation (DMC) owns two mines in Pennsylvania. The coal from the mines is separated into two grades before shipping. The Pittston mine produces an average of 10 tons of low-grade and 15 tons of high-grade coal daily at an operating cost of $10,000. The Altoona mine produces an average of 12 tons of low-grade and 8 tons of high-grade coal daily at a cost of $6200. Contractual obligations require DMC to ship 500 tons of low-grade and 600 tons of high-grade coal each month. Determine the number of days each mine must operate to meet the company's contracts at a minimum cost.

21) A bag of dog food manufactured by Fido, Inc., must meet the guaranteed analysis printed on the bag: At least 14% crude protein, not less than 8% crude fat, at most 5.5% crude fiber and no more than 12% moisture. There are three ingredients that can be mixed to produce the tasty dog food that dogs prefer: I (yellow corn, bran) costs $0.45 per pound and contains 12% crude protein, 2% crude fat, 11% crude fiber and 20% moisture. II (chicken, beef tallow) costs $1.07 per pound and contains 26% crude protein, 31% crude fat, 1% crude fiber, and 8% moisture. III (ground wheat and brewer's rice) costs $0.60 per pound and contains 10% crude protein, 0% crude fat, 18% crude fiber and 6% moisture. What mix of the three ingredients -- I, II, and III -- should be mixed to total one pound for a minimum cost?

22) Two warehouses have canned soup on hand and three stores require soup in stock. UFT, United Food Transport, has contracted to deliver the soup. Warehouse 1 has 125 cases on hand and Warehouse 2 has 175 cases. The stores which we will call A, B, and C, all have needs of 100 cases each. The costs per case for shipping between the warehouses and stores are given in the table below. Find the solution that minimizes the cost of shipping.

	store A	store B	store C
warehouse 1	$0.08	$0.10	$0.28
warehouse 2	$0.10	$0.18	$0.15

23) Metalheads, Inc., wishes to start production and sale of a new metal alloy containing iron, copper, and lead. The alloys containing iron, copper, and lead available to buy are A, B, C, D, and E. The percent of iron, copper, and lead in each and in the desired alloy are:

	A	B	C	D	E	needed .
% iron	10	10	40	60	30	exactly 30
% copper	10	30	50	30	40	exactly 30
% lead	80	60	10	10	30	exactly 40
cost/lb	3.9	4.1	6.0	6.03	7.7	

What is the best combination of the alloys A, B, C, D, and E to give exactly the needed percentages that will minimize the cost?

SOLVING PROBLEMS

2

- Systems of Linear Inequalities --

A linear programming problem that has two variables can be solved graphically. The bad news is that solving by hand takes several steps, but the good news is that the steps are the same for any problem. First we will graph the feasible region by graphing all of the constraints, then find a few suitable points that are candidates for the best one, and choose among them by evaluating the objective function. We review graphing first.

The Cartesian Coordinate System

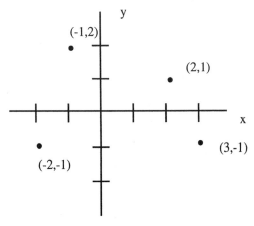

We graph lines and inequalities using the Cartesian coordinate system. Points on the graph are identified with a pair of numbers, the first number or coordinate being the distance to go right or left and the second the distance to go up or down. On this coordinate system those directions are called x and y, but they will be named with your variable names when solving linear programming problems. (2,1) means 2 units to the right and 1 unit up. (-2,-1) means 2 units to the left and 2 units down, etc.

Always label the axes with the variable names, as we did with x and y in the picture. Mathematical convention says that the horizontal axis variable comes first in the ordered pair with the vertical axis second, so the point is always read left to right as horizontal movement, vertical movement. You should follow this convention whenever you graph ordered pairs.

Graphing Inequalities

The constraints in linear programming usually have the form $ax + by \leq c$, where x and y are variables, and a, b, and c are constants. We start by graphing the associated line: $ax + by = c$.

Example 2.1: Let's graph $0.2s + 0.5m = 3$.

Solution: From geometry we know that two points determine a line We will use the points where one of the variables is zero:

If $s = 0$ then $0.5m = 3$ so dividing both sides by 0.5 we get $m = 6$.

If $m = 0$ then $0.2s = 3$ so dividing both sides by 0.2 we get $s = 15$. In table form.

$$0.2s + 0.5m = 3$$

s	m
0	6
15	0

Now that we have the points that we need we plot them and draw in the line. It helps to use graph paper for this. We start by drawing the axes and choosing a scale. The variables are called s and m so we have labeled the axes accordingly. The negative parts of the graph are ignored because of the non-negativity constraints we will have in all of our problems.

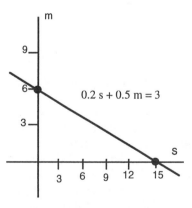

Next we plot the points. Having labeled the horizontal axis s, we interpret the first number as the s-coordinate or horizontal movement. Similarly the second number is the m-coordinate or the vertical movement. To plot (0,6) we move 0 horizontally and 6 vertically. Plot (15,0) and connect the points.

YOU TRY IT 2.1

Plot $0.2s + 0.3m = 1$.
Be sure to show what
scale you are using.

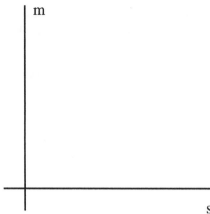

Now we move on to graphing linear inequalities. We start by drawing the associated line (replace the inequality with an equality.) Suppose that the line we graphed in Example 2, $0.2s + 0.5m = 3$, actually came from the inequality constraint $0.2s + 0.5m \leq 3$. Let's experiment to discover which points satisfy the inequality and which ones don't.

Point	$0.2s + 0.5m \leq 3$	where
(2,6)	false	above
(3,8)	false	above
(5,2)	true	on the line
(10,1)	true	below
(1,10)	false	above
(0,0)	true	below
(100,100)	false	above

Where are the "true" points? Look back at the graph of the line and plot them. They're either below the line or one it. So the points that work are on the line and one side and those that don't are on the other side. This will always be the case. **The solution to a linear inequality is the line and one side of it;** all the points on one side of the line satisfy the inequality and all the points on the other side don't. Of course the points on the line satisfy the equality so they work for $\geq$ or for $\leq$.

To graph a linear inequality first graph the line, then pick one point not on the line and try it in the inequality. If that point makes the inequality true the side that point is on (plus the line itself) is the solution to the inequality, if the point makes the inequality false the other side is the solution. **We will mark out the side that doesn't work by shading it.**

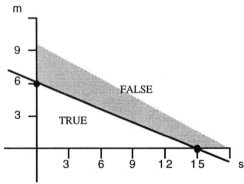

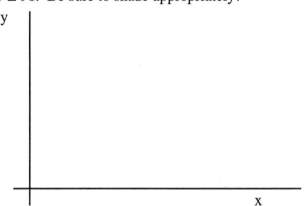

YOU TRY IT 2.2

Graph 9x + 15y ≥ 90. Be sure to shade appropriately!

Graphical Solution

Now we will tackle the Barbie and Ken problem graphically and then give an outline for solving these problems in general. The constraints are:

```
(plastic)          12B + 14K ≤ 100,000
(cardboard )       4B + 4K ≤ 35,000
(nylon)            5B ≤ 30,000
(sales)            B - 2K ≥ 0
(non-negativity)   B ≥ 0 and K ≥ 0.
```

We will graph B ≥ 0 and K ≥ 0 by using only the positive parts of the axes. To graph the others we will make a table for each one, look at the tables to see what scale to use on our graph, graph the lines and choose the correct side for each inequality, shading the side that **doesn't** work; **the region with no shading will be the feasible region!**

The first two inequalities are very similar to the one you just tried. We plot them and then check the shading by testing (0,0) in each.

Inequality:	12B + 14K ≤ 100,000		4B + 4K ≤ 35,000	
Chart:	B	K	B	K
	0	7142.86	0	8750
	8333.3	0	8750	0

Test Point: (0,0) True: shade away   (0,0) True; shade away

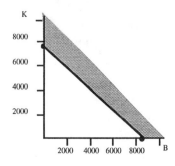
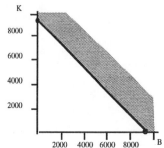

The other two constraints are special. The first (nylon) has only one variable. This type of equation can be solved for the value of the only variable, in this case B = 6000. **Any** value for the missing variable will work, so choose two of your favorites (say 0 and 1000). Plot the points and check shading with (0,0).

The second (demand) has a zero right hand side, so we get the same point, (0,0), when plugging 0 in for the variables whereas before we got two. We choose another value for one of the variables arbitrarily (say 6000), staying near the other values in the problem, and calculate K (3000). Since (0,0) is on the line, it gives no information about shading. We arbitrarily test (0,4000) for shading.

Inequality:	$5B \leq 30,000$		$B - 2K \geq 0$	
Chart:	B	K	B	K
	6000	0 (arbitrary)	0	0
	6000	1000 (arbitrary)	6000 (arbitrary)	3000

Test Point: (0,0) True; shade away   (0,4000) False; shade towards

We have graphed all these lines separately, but we really need to graph them all on one coordinate system to see the feasible region. Remember that the feasible region will appear in our graph as the only region with **no** shading!

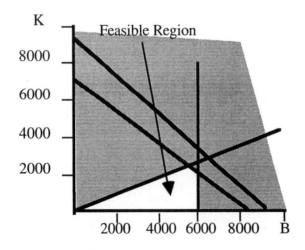

240

YOU TRY IT 2.3
To familiarize yourself with the feasible region find three points in it and compute the profit for each point. We have done one to get you started. Remember that $Z = 6B + 6.5K$.

POINT(B, K)	PROFIT(6B+6.5K)
(5000,1000)	$36,500.00
_____	_____
_____	_____
_____	_____

Now that we have the feasible region graphed, we want to find the best point, but it is clear that there are too many points to try. We will need some way of choosing the "best" candidates for the maximum. We decided in Section 1 that the best answer must occur when some of the constraints are pushed to the limit. What does this mean on the graph?

Let's look at the plastic constraint. There's no room left in this constraint when $12B + 14K = 100,000$. Graphically, this happens on the line we drew for the plastic constraint. So where on the graph will two constraints have no room left? When two of the lines we drew cross! In other words at a corner of the region. This is the fundamental idea for the method we'll use to solve the problems. We'll use the important mathematical property:

> **The best solution will always occur at a corner point of the feasible region.**

If we can find the best point among the corner points we will have the best point in the entire feasible region. That means we have reduced the problem from trying an infinite number of points to only trying the corners, four in this case. Of course by best we mean the point that makes us the most profit. Let's list the corner points for Barbie and Ken, and see which corner gives the best profit.

Corner Point (B,K)	Profit (6B + 6.5K)
(0,0)	$0
(6000,0)	$36,000
(6000,2000)	$49,000
(5263.16,2631.58)	$48,684.23

Since the third corner point gives the biggest profit, the theorem says that (6000,2000) must give the best or **optimum** solution for all the points. So for the Barbie and Ken problem, the very best we can do is to make 6000 Barbie dolls and 2000 Ken dolls for a total profit of $49,000. Here is a summary of the steps we followed:

Solving a Linear Programming Problem using Graphing
1. Build a table of values for the line associated with each inequality constraint.
2. Draw the axes, using a scale that will work for all the constraints, and then graph each of the lines.
3. Shade the "throw away" side for each linear inequality. The feasible region is the region with no shading!
4. Find all the corner points of the feasible region.
5. Test the objective function at each of the corner points. Choose the best from this list; the largest objective is the maximum, the smallest is the minimum.

So by now you must be wondering how to find the corner points. What exactly is a corner point? We know what it is geometrically, but what is it algebraically? Look below at the feasible region for the Barbie and Ken problem again with the corner points numbered. They occur where lines associated with the inequalities cross! In other words each corner is at an intersection of two lines. Intersection points we can solve for algebraically. Let's find them.

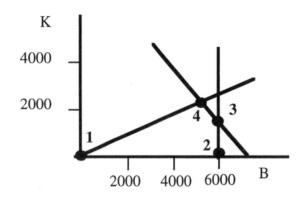

1) $B=0$ and $K=0$ intersection: Clearly the point is $(0,0)$.

2) $K=0$ and $5B=30,000$ intersection: Again this is clearly $(6000,0)$ from the work we did while graphing. (Be careful! The B-coordinate comes first.)

3) $5B=30,000$ and $12B + 14K =100,000$ intersection: Here we need to solve the equations $5B = 30000$ and $12B + 14K = 100,000$ simultaneously. Let's solve the first equation for B and substitute the result into the second equation:

$$5B = 30,000 \text{ gives } B = 6,000 \qquad 12B + 14K = 100,000$$
$$12(6000) + 14K = 100,000$$
$$72,000 + 14K = 100,000$$
$$K = 2000$$

So the point is $(6000,2000)$.

4) $12B + 14K = 100,000$ and $B - 2K = 0$ intersection: Again we'll solve one of the equations, this time the second, for a variable and substitute the result into the other.

$$B - 2K = 0 \quad \text{gives} \quad B = 2K \qquad 12(2K) + 14K = 100000$$
$$K = 2631.58 \text{ (approximately)}$$

Substituting that back into $B = 2K$ gives $B = 2(2631.58) = 5263.16$.
So the point is (5263.16, 2631.58).

Note that there is another intersection of two constraints on the graph, directly above corner point 3, but it is **not** a corner we need to check because it is outside the feasible region (remember the shading!) so we ignore it. Let's practice our intersection finding skills with some more examples.

Example 2.2: Find the point of intersection of the two lines

$$3x + 5y = 23 \text{ and } x - y = 1.$$

Solution: It looks simpler to solve for one of the variables in the second equation, so we solve for x in that equation and substitute into the first equation.

$$x - y = 1 \quad \text{gives} \quad x = y + 1 \qquad 3(y+1) + 5y = 23$$
$$3y + 3 + 5y = 23$$
$$8y + 3 = 23$$
$$y = 2.5$$

Then $x = y + 1$, so $x = y + 2.5 = 3.5$ The solution then is $x = 3.5$ and $y = 2.5$.

This method of solving 2 equations in 2 unknowns is called solving by **substitution**. There is another way that is sometimes faster. Suppose that we needed to find the intersection of

$$2x + 3y = 24 \text{ and } 5x + 9y = 66.$$

We could solve for x or y and substitute as before, but that would involve fractions. It's easier in this case to add together multiples of the equations to **eliminate** one of the variables and solve for the one that remains.

We eliminate x by multiplying **both sides** of the equations appropriately so that when we subtract them we get 0x:

```
5(2x + 3y = 24)   gives      10x + 15y = 120
2(5x + 9y = 66)   gives      10x + 18y = 132

Now subtract to get                 -3y = -12

and divide by -3 to get               y = 4
```

Now that we have one of the variables, we proceed the same way as before: Substitute back into one of the original equations:

```
                                  2x + 3y = 24
                                  2x + 12 = 24
                                       2x = 12
                                        x = 6
```

So the point of intersection is (6,4). This is called solving by **elimination**.

Whichever method you like to use, you can be sure that you've found the intersection by checking that the point works in both of the original equations!

```
2(6) + 3(4) = 12 + 12 = 24 (check)
5(6) + 9(4) = 30 + 36 = 66 (check)
```

YOU TRY IT 2.4

a) Find the point of intersection of the lines $2x = 15$ and $3x + 5y = 40$.

b) Find the point of intersection of the lines $3d - 7p = 22$ and $11d + 4p = 8$.

Now we are ready to graphically solve a new linear programming problem from beginning to end. Remember to follow the steps carefully. They are repeated here for you.

	Solving a Linear Programming Problem using Graphing
1.	Build a table of values for the line associated with each inequality constraint.
2.	Draw the axes, using a scale that will work for all the constraints, and then graph each of the lines.
3.	Shade the "throw away" side for each linear inequality. The feasible region is the region with no shading.
4.	Find all the corner points of the feasible region.
5.	Test the objective function at each of the corner points. Choose the best from this list; the largest objective is the maximum, the smallest is the minimum.

Example 2.3: The manufacturing process at a particular oil refinery requires it to produce at least 2 gallons of gasoline for each gallon of fuel oil. To meet the winter demand for fuel oil, at least 3,000,000 gallons a day must be produced. The demand for gasoline is no more than 8 million gallons a day. If the refinery makes $0.88 for each gallon of gas and $0.65 for each gallon of fuel oil, how much of each should be made to maximize profit?

Solution: This problem was translated in Example 1.1 in Section 4.1. The mathematical formulation is repeated below. Notice that the quantities have been scaled slightly to remove the very large numbers from the calculations.

```
G = the number of gallons of gas (in millions)
F = the number of gallons of fuel oil (in millions)

Maximize Z = 650,000 F + 880,000 G
      Subject to:
        (1) sales:      2 F -  G ≤ 0
        (2) oil demand:   F       ≥ 3 (in millions)
        (3) gas demand:        G ≤ 8 (in millions)
                     F ≥ 0, G ≥ 0
```

The feasible region is:

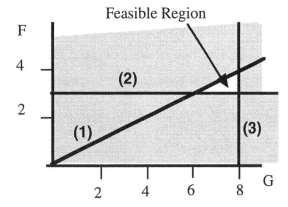

Feasible Region

The corner points:

F = 3 intersects G - 2F = 0. Solving the equations using substitution:

$$G - 2(3) = 0$$
$$G = 6$$

Point: (6, 3).

$F = 3$ intersects $G = 8$.
Point (8,3).

$G = 8$ intersects $G - 2F = 0$. Solving with substitution:
$$(8) - 2F = 0$$
$$8 = 2F$$
$$F = 4$$

Point: (8, 4).

Putting this into a table and evaluating the objective function at these corner points gives:

POINT	PROFIT
(G,F)	880,000G + 650,000F
(6,3)	7,230,000
(8,3)	8,990,000
(8,4)	9,640,000

Optimum: 8 million gallons of gasoline and 4 million gallons of fuel oil give a profit of $9,640,000. Not a bad day's work.

Let's work through one more problem completely. As you read the solution, think about how each step in the process of translation and graphical solution is accomplished.

Example 2.4: A chemical plant needs to cut its emission of sulphur dioxide (SO_2) and suspended particulate matter to satisfy federal regulations. It can use scrubbers in the smokestack or filters in the process. A filter can reduce particulate matter by 12 pounds per hour and SO_2 emissions by 2 pounds per hour. A scrubber reduces particulate matter by 7 pounds per hour and SO_2 emissions by 4 pounds per hour. Filter units cost $4000 and scrubber units cost $6000. The smokestack can hold only 12 scrubbers and no more than 10 filter units can be installed. How can the plant reduce SO_2 emissions by at least 44 pounds per hour and particulate emissions by at least 84 pounds per hour at least cost?

Solution: The linear programming translation is as follows.

```
F = the number of filters to install
S = the number of scrubbers to install
```

Then we are asked to

```
minimize 4000F + 6000S
subject to
        (1)  SO2              2F + 4S ≥ 44
        (2)  Particulate     12F + 7S ≥ 84
        (3)  Max scrubbers         S ≤ 12
        (4)  Max Filters      F        ≤ 10
```

Now we graph the inequalities and find the feasible region.

First we make tables for each inequality turned into an equation:

2F + 4S = 44		12F + 7S = 84		S = 12		F = 10	
F	S	F	S	F	S	F	S
0	11	0	12	0	12	10	0
22	0	7	0	12	12	10	12

Then we graph each of the lines, check which side of the line makes the inequality true, shading the side that isn't feasible. That leaves us with the feasible region.

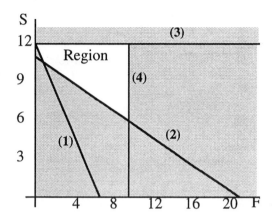

$2F + 4S = 44$ and $12F + 7S = 84$ intersect in a point that is a corner point of the region. To find the point using elimination, we multiply the second equation by 4, the first by 7 and subtract.

$$
\begin{array}{rcr}
48F + 28S & = & 336 \\
- \ 14F - 28S & = & -308 \\
\hline
34F & = & 28 \\
F & = & .8235
\end{array}
$$

Substituting this in:
$$2(0.8235) + 4S = 44$$
$$4S = 44 - 1.647 = 42.353$$
$$S = 10.59$$

The other points are easy to find from the charts we used to draw the graph. Here they are with the costs for each.

point (F,S)	cost = 4000F + 6000S	
(0.8235, 10.59)	$66,834	**BEST**
(0,12)	$72,000	
(10,12)	$112,000	
(10, 6)	$76,000	

So the best thing to do is to install 0.8235 filters and 10.59 scrubbers, assuming the units can be made in these amounts.

YOU TRY IT 2.5

Translate and solve this linear programming problem graphically, following the 5 steps given above. Be sure to list the math formulation, graph the feasible region and fill in the table below with the corner points. Write in a sentence what Sally should do and how much she will make if she does.

Sally is trying a new line of vacuum cleaners in her store. The line has two models, the Supervac and the Econovac. She has 61 square feet of floor space to devote to this line. An Econovac takes up 3 square feet of space and a Supervac 4. Each Supervac requires an investment of $150.00 and each Econovac $80.00. Sally only wants to invest $2060 in the product line, and she estimates that for each model she displays for a year she will sell one, making $30.00 for each Econovac sold and $50.00 for each Supervac. How many of each should she display to maximize her profit?

Math Formulation:

Graph:

	CORNER POINT	PROFIT
	_____	_____
	_____	_____
	_____	_____
	_____	_____

feasible region

Answer: *Sally should stock* _____

In all of the examples so far, the feasible region has been a closed polygon, so the number and approximate location of the corner points was immediately obvious from the graph. Sometimes the feasible region seems to be missing a side, as in the example below. This region is referred to as **unbounded**. We can think of the missing corners as being infinite. Such problems can still have solutions.

Example 2.5: The Watauga county refuse department runs two recycling centers. Center 1 costs $40 to run for an eight hour day. In a typical day 140 pounds of glass and 60 pounds of aluminum are deposited at Center 1. Center 2 costs $50 for an eight hour day, with 100 pounds of glass and 180 pounds of aluminum deposited per day. The county has committed to deliver at least 1540 pounds of glass and 1440 pounds of aluminum per week to encourage a recycler to open up a plant in town. How many days per week should the county open each center to minimize it's cost and still meet the recycler's needs?

Solution: The linear programming translation is as follows.

```
c1 = number of days to open center 1
c2 = number of days to open center 2
```

Then we are asked to

```
minimize    40 c1 +  50 c2 (cost)
subject to
            140 c1 + 100 c2 ≥ 1540 (glass)
             60 c1 + 180 c2 ≥ 1440 (aluminum)
```

The tables of points for these two inequalities are:

$140 c_1 + 100 c_2 = 1540$			$60 c_1 + 180 c_2 = 1440$	
c1	c2		c1	c2
0	15.4		0	8
11	0		24	0

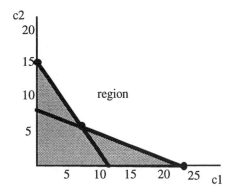

The region is unbounded, with three finite corners. Two are easy from the graph: (0, 15.4), (24, 0). The third must be found (verify): (6.94, 5.69). Note that the "infinite" corners cannot be best since we're minimizing.

point (c1,c2)	cost $= 40 c_1 + 50 c_2$	
(0, 15.4)	$770	
(24,0)	$960	
(6.94,5.69)	$562.10	**BEST**

Open Center 1 for 6.94 days per week and Center 2 for 5.69 days. The fractional parts can be interpreted as less than eight hour days: 6.94 days gives six full days and one 7.5 hour day; 5.69 days gives five full days and one 5.5 hour day.

YOU TRY IT 2.6

Solve the following problem graphically. (You translated it in YTI 1.2 in section 1.)

A dietitian wishes to prepare a meal which has a minimum of calories but still satisfies nutritional requirements. In particular it must have at least 3 oz of protein. 1 oz of fat, and 3 oz of carbohydrates. The foods available are: a salad which has 30 cal/oz and is 20% protein, 20% fat, and 60% carbohydrates; and meat which has 70 cal/oz, is 50% protein, 30% fat, and 20% carbohydrates. What amounts of salad and meat should she serve?

One last word on graphical solutions

Several of the problems formulated in Section 1 had more than two variables.. Such a problem is difficult to examine graphically. If there were three variables, we could try to draw a 3D graph; more than three variables means drawing a region to examine would be impossible. In the next section we will learn how geometric ideas can be translated into algebraic ideas and then implemented on a computer to arrive at a solution.

Exercise Set 4.2

1) Graph $3x + 5y \geq 20$.

2) Graph $5x - 3y \leq 30$.

3) Graph $3x \leq 12$.

4) Graph $7y \geq 10$.

5) Suppose this is the feasible region for a linear programming problem with objective $Z = 5x + 3y$. What is the maximum value of the objective in this region? The minimum?

6) Suppose this is the feasible region for a linear programming problem with objective $Z = 100 R + 223 T$. What is the maximum value of the objective in this region? The minimum?

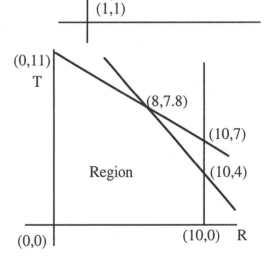

In 7 through 18 formulate the problem and find a graphical solution.

7) (Problem 5 from Section 1) A chocolate bar contains 50 calories and 5 grams of protein, and makes a child smile 7 times while eating it. A granola bar contains 25 calories and 15 grams of protein, and makes a child smile 2 times while eating it. A teacher plans to serve some of these snacks to his first grade class. They should total at least 40 grams of protein and no more than 150 calories. How many of each kind of bar should he serve to maximize smiles?

8) (Problem 6 from Section 1) Sarah has decided to try to make some money by making and selling mailboxes. She has contacted a store that will buy all she can produce. She makes regular mailboxes and deluxe mailboxes, and

makes a profit of $9 on the regular mailboxes and $16 on the deluxe mailboxes. It takes an hour to make a regular mailbox and 3 hours to make a deluxe mailbox. Her work area is so small that she can only paint 7 total mailboxes a day. The store wants at least 3 mailboxes each day. Sarah works hard, but a 10 hour workday is about all she can manage. How many of each type of mailbox should she make to maximize profit?

9) (Problem 7 from Section 1) Mary wants to sell fruit punch and lemonade on the sidewalk in front of her house. In conference with her mom, she has made the following decisions:

Selling price: Fruit punch at 15 cents per cup

Lemonade at 12 cents per cup

In addition, Mary has estimated that her cost to provide a cup of fruit punch is 5 cents per cup and to provide a cup of lemonade is 3 cents per cup. Mary has a total of 40 cups for her operation, and available funds of $1.50. How many of each type of drink should she sell to maximize profit?

10) (Problem 8 from Section 1) A manufacturer makes 2 kinds of cars, the sedan model and the sports model. According to a contract, it must supply at least 5000 cars. The sports model has a special engine of which only 2500 are available. Both models are made on an assembly line on which only 15,000 hours of time are available. The sedan takes 2 hours on the line and the sports car 4 hours. The profit is $750 on the sedan and $2000 on the sports model. How many of each should be made to maximize profit?

11) (Problem 9 from Section 1) The Boone Weavers make shawls and afghans. They spin yarn, dye yarn and weave yarn for each. A shawl requires 1 hour of spinning, 1 hour of dyeing and 1 hour of weaving. An afghan requires 2 hours of spinning, 1 hour of dyeing and 4 hours of weaving. In a week, there is time to spend at most 13 hours spinning, 10 hours of dyeing and 30 hours weaving. How many shawls and afghans should be made to maximize profit if shawls bring in a profit of $25 and afghans a profit of $40?

12) (Problem 10 from Section 1) Joe's Bottlers sells cola and uncola drinks in vending machines. Each machine will hold at most 150 bottles. A bottle of cola brings in $0.25 profit, and a bottle of uncola $0.45. Since more colas than uncolas are sold, they want at least twice as many colas as uncolas, but there should be at least 20 of each kind in a machine. How many of each should be in a machine to maximize profit?

13) (Problem 11 from Section 1) An office manager needs to purchase storage lockers. She knows that a regular locker costs $40, requires 4 square feet of floor space, and holds 10 cubic feet of materials. On the other hand, a large storage locker costs $100, requires 8 square feet of floor space, and holds 20 cubic feet. Her budget permits her to spend at most $600, while the office has room for no more than 52 square feet of lockers. The manager desires the greatest cubic feet of storage capacity within the limitations imposed by funds and space. How many of each type of locker should she buy?

14) (Problem 14 from Section 1) A textile plant processes 2 types of cloth, an acrylic-cotton (AC) blend and a nylon-cotton (NC) blend. Each is blended 50-50. Acrylic and nylon are available in unlimited supply, but due to a recent drought, only 60000 pounds of cotton are available each month. The spinning department requires 1 hour of production time per 1000 pounds of

AC blend and 2 hours per 1000 pounds of NC blend. This department has 150 available hours per month. The dye room has 150 available hours per month. The AC blend requires 0.5 hours per 1000 pounds in the dye room and the NC blend requires 3 hours. If the profit from the AC blend is $40 per 1000 pounds and $50 per 1000 pounds of the NC blend, how much of each blend should be produced in a month to maximize profits?

15) (Problem 17 from Section 1) Mary, who is ill, takes mineral supplements. Each day she must have at least 21 units of iron, 10 units of potassium, and 40 units of calcium. She can choose between pill #1 which contains 7 units of iron, 2 of potassium, and 4 of calcium, and pill #2 which contains 3 units of iron, 2 of potassium, and 14 of calcium. Pill #1 costs $0.20 and pill #2 costs $0.35. Find the number of each pill which would minimize cost for Mary.

16) (Problem 18 from Section 1) A dietary manager at a retirement center plans to serve a lunch of roast beef and bread. She is required by law to serve at least 400 calories and 3 milligrams of iron, but would like to minimize her cost. She can buy roast beef for 8¢ per ounce (100 calories and .8 milligrams of iron) and bread for 1¢ per slice (25 calories and .1 milligram iron). Find the amounts of roast beef and bread she should serve.

17) (Problem 19 from Section 1) A small generator burns two types of fuel, low sulfur and high sulfur, to produce electricity. In one hour of use each kilogram of low sulfur fuel emits 3 units of sulfur dioxide, generates 3 kilowatts and costs 75 cents, while each kilogram of high sulfur fuel emits 6 units of sulfur dioxide, generates 3 kilowatts and costs 50 cents. The EPA insists that the maximum amount of sulfur dioxide emitted per hour be 9 units. At least 12 kilowatts must be generated per hour. How many kilograms of each fuel should be used hourly to minimize the cost of the fuel used?

18) A trucker delivers 30-pound and 50-pound concrete blocks to a construction company. His contract calls for the delivery of at least 10,000 blocks weighing a total of at least 200 tons each week. It costs the trucker 8¢ to deliver a 30-pounds block and 10¢ to deliver a 50-pound block. How can he minimize cost?

19) An investor has $50,000 to invest. Option A yields a return of 15% per year and has a risk factor of 25 (the higher the risk factor, the higher the possibility of a loss). Option B yields a return of 10% per year and has a risk factor of 15. How much should the investor invest in each option if he wants to achieve at least a 13% return and minimize risk?

20) In the United States the number of miles driven each year is approximately $1.872*10^{12}$. The amount of carbon dioxide (CO_2) emitted is $1.920*10^{12}$ pounds. By using methanol to replace more of the gasoline we can eliminate some of the CO_2 and thus reduce pollution from this gas. Gasoline emits 20 lb. of CO_2 per gallon and methanol emits 9.5 pounds of CO_2 per gallon. We will require that CO_2 emissions be dropped to $1.8*10^{12}$ pounds per year. Cars on average get 19.4 miles per gallon of gasoline but get only 11 miles per gallon on methanol. Gas is $1.20 per gallon and methanol is $.70 per gallon and we insist that we spend no more than $120*10^9$ on fuel.

a) Compute the cost per mile of driving using each fuel and the number of pounds of CO_2 produced per mile for each fuel.

b) Find the maximum number of miles that could be driven given these constraints.

c) If methanol were not available, how far could cars be driven subject to these constraints?

SOLVING LARGE PROBLEMS AND SENSITIVITY

3

Solving a two variable linear programming problem graphically is straightforward. It is also possible to solve a three variable problem graphically, but the corner points become more difficult to locate visually. It is impossible to solve a problem with more than three variables using our graphical method. We have already translated problems with three or more variables; how do we solve them? The corner points of the feasible region must be found somehow using algebra and then checked for the optimum value. The **Simplex** algorithm developed by Danzig solves the problem using this idea.

The simplex method starts with any **corner** of the feasible region, usually the origin, (did you notice that (0,0) is often one of our corners?) and then tests to see which edge points towards the neighboring corner with the biggest change for the better in the objective. Then the method "walks" along that edge to the new corner. At the new corner the edges are tested again. If none give a change for the better in the objective, this new corner is the answer, otherwise the method walks and tests again.

In this manner the simplex method walks along the edges to the best corner. To explain this in detail requires some fairly sophisticated algebra that we won't discuss here. The important thing is that there are many computer programs that will solve a linear programming problem quickly, even problems with more than 100 variables.

The rest of this section assumes that you have access to such a package, and that you have learned how to enter the mathematical formulation of the problem and get a solution.

YOU TRY IT 3.1

Enter the Barbie and Ken problem (see page 197) into your linear programming package and verify the answer from the graphical solution. (K = 2,000 and B = 6,000 giving a profit of $49,000)

Having access to a computer program to solve these problems allows us to answer "what if" questions easily. Let's consider the following new example.

Example 3.1: Swales Jeweler's uses diamonds and emeralds to produce two types of necklaces. One requires 2 diamonds, 3 emeralds, 1 hour of labor and brings in a profit of $200. The second requires 3 diamonds, 2 emeralds and 2 hours of labor and Swales makes $250 on each one sold. Swales has 100 diamonds, 120 emeralds and 70 hours of labor available each week. Market demand requires that at least 24 of the second type necklace be produced.

a) How many of each type of necklace should be produced to maximize profit?

b) How does the answer change if the profit from the type one necklace falls to $180?

c) How does the answer change if instead the profit from the type one necklace rises to $235?

Solution: Clearly one variable should be the number of type 1 necklaces to make and another should be the number of type 2 necklaces to make.

```
N1 = the number of type one necklaces to make.
N2 = the number of type two necklaces to make.
```

The objective is straight forward: $Z = 200N1 + 250N2$

The chart for the constraints looks like

	N1	N2	limit
Diamonds	2	3	100
Emeralds	3	2	120
Labor	1	2	70
N2 Demand		1	24

This gives the following math formulation:

```
Let N1 = the number of type one necklaces to make
    N2 = the number of type two necklaces to make

maximize           Z = 200N1 + 250N2
subject to diamonds:    2N1 +   3N2  ≤ 100
           emeralds:    3N1 +   2N2  ≤ 120
           labor:        N1 +   2N2  ≤ 70
           N2 demand:            N2   ≥ 24
                       N1 ≥ 0   N2 ≥ 0
```

Now we use this simple example to try out the linear programming software. Here is the solution for part a:

```
Z = 8,800    N1 = 14    N2 = 24
```

So we should make 14 of the first type necklace and 24 of the second. Profit will be $8,800.

Now we can answer the "what if" questions in parts b and c: Simply change the 200 in front of the N1 in the objective function first to 180 and then to 235 and the answer changes to:

```
Profit from N1 = $180        Profit from N1 = $235
        Z = 8,520                    Z = 9,290
        N1 = 14                      N1 = 14
        N2 = 24                      N2 = 24
```

> Notice that the profit changed, (decreased when the price fell and increased when the price went up -- can you explain why?) but the actual numbers of necklaces made did not change! So a small change in the profit from the type one necklace has little impact on how many of each necklace to make for the best profit.

We will see that the reverse can happen as well; changing a number even by a very small amount will completely change the entire solution. If this happens we say the problem is **sensitive** to that change.

This example could have be done graphically by hand, even the "what if" questions. The questions asked dealt with changes to the objective, so the graph of the feasible region wasn't affected. If the questions we ask deal with changing a constraint, the feasible region may change substantially, so we'd have to re-draw the region and recalculate some of the corners. That's where the power of a computer program really counts.

YOU TRY IT 3.2

Answer the following questions about the Barbie and Ken problem.
a) How does the answer change if the profit from the Ken doll is raised to $6.95?
b) How does the answer change if the weekly shipment of plastic is increased to 150,000 ounces?
Explain what you did in each case to check.

Slack Variables

Most computer programs will tell you more than just the answer to your linear programming problem. Often included are the **slack** variables for each constraint. These numbers will tell you how much is left in the constraints at the optimal solution. A slack variable is usually labeled with an **S** followed by the number of the constraint (given the order the constraints were entered into the program).

For the original necklace problem of Example 3.1, the slack variables are:

$$s1 = 0.0$$
$$s2 = 30.0$$
$$s3 = 8.0$$
$$s4 = 0.0$$

S1 is the amount left over from the first inequality. This was the diamond constraint, and we don't have any left over since we used $2(14) = 28$ for the type one necklace and $3(24) = 72$ for the second, to total 100 (the number we had). S2 is the number of extra emeralds we have. We started with 120 emeralds, used $3(14) = 42$ to make the first kind of necklace, used $2(24) = 48$ to make the second kind of necklace for a total of 90 emeralds used. So 30 are left over. S3 is the left over labor. S4 is the "left over demand", i.e., we exactly matched the demand. So we can say:

The slack variables tell us how much is left over in the corresponding constraints.

YOU TRY IT 3.3

Verify that the slack for the labor constraint in the necklace problem is correct by calculating it using the number of necklaces made.

Shadow Prices

Depending on how advanced your software is, it may give you much more than just the optimal values of the variables, objective and slacks. It may also tell you ways to make more money, how much the constraints are costing, and how much the situation can change without making a considerable difference in the solution.

One question we often ask is "how valuable is a resource or raw material?" This leads to questions like "If we had room or funds for a little more of one of the resources, which one should we get?" The unit worth or value of one unit of a resource is called a **shadow price** or sometimes a dual price in the program printout. To find the shadow prices, you will need to check out where they are given by your program.

Let's consider the shadow prices from the necklace problem solved in Example 3.1:

Constraint	Shadow Price
Diamond	100.0
Emerald	0.0
Labor	0.0
N2 Demand	-50.0

1) The shadow price for the diamonds is 100.00; another diamond would be worth $100.00 in profits to us if we could get more of them. If we had one less diamond we would lose $100 in profits.

2) The shadow price for the emeralds is 0.0. We didn't use them all anyway (see slack variable S2), so it makes sense that we wouldn't place any value at all on having more of them. It doesn't mean that the emeralds aren't important to us; it means that having **more** of them isn't.

3) From the labor shadow price of 0.0 we see that extra labor won't help. This makes sense since we know we have 8 hours left over. (See S3.)

4) The -50.00 shadow price for N2 demand means that if demand increased by 1, it would **cost** us $50.00; decreased by 1, it would save us $50.00.

A shadow price tells us how much the objective changes if we change the right-hand-side of one of the constraints by one.

YOU TRY IT 3.4

Now you will see if the shadow prices really do what we have said that they do. Solve the necklace problem changing the three constraints below by increasing the right-hand-sides by one and then decreasing them by one. In each case record what happens to the objective function. DO EACH ONE SINGLY! Compare how much more or less profit there is compared to the amount that is predicted by the shadow prices.

	PROFIT	
Inequality	Increase by one	Decrease by one
Diamonds		
Labor		
Emeralds		

Note: These numbers only give us information when the values of the variables are at (or very close to) the optimal values. Soon we will find out how far from the solution you can go and still have these numbers hold.

Sensitivity Ranges

The other questions we might like to ask have the form "what happens if XXX changes?" We've already seen that we can make the changes to the mathematical formulation of the problem and rerun the program. When there are hundreds of variables in the problem, we don't want to rerun the program unless we absolutely have to. In certain situations we don't. Limits on how much we can change some of the numbers in the problem without changing the final answer are referred to as **sensitivity ranges**. They are usually summarized in a table in the computer program's output. Ranges are given for the right-hand-sides of the constraints and the coefficients in the objective function. Here are the sensitivity ranges for the original version of the necklace problem from Example 3.1:

	right-hand-side ranges	
constraint	lower	upper
diamonds	72.0	116.0
emeralds	90.0	+ infinity
labor	62.0	+ infinity
N2 demand	12.0	33.33

	objective coefficient ranges	
variable	lower	upper
N1	166.67	+ infinity
N2	− infinity	300.0

The first set of ranges are for the right-hand-sides of the constraints. Each has a lower limit and an upper limit. In between those limits the shadow prices remain constant.

1) Notice what a narrow band the diamond constraint has. That means that the shadow price will change if we have less than 72 diamonds and if we have more than 116 diamonds. We would expect the shadow price to increase if we had less than 72 diamonds and decrease if we had more than 116 since scarce resources are more costly than plentiful ones.

2) We could go down to 90 emeralds and still have 0 as the shadow price, and of course if we got more of them the shadow price would remain 0. Notice that 90 is how many emeralds we used in the optimal solution.

3) The range of 12 to 33.33 for the N2 demand constraint means that if we required production of more than 33.33 of the type 2 rings the shadow price would go up, and we could drop production to 12 per week before the shadow price for that inequality dropped.

> **The right-hand-side ranges tell us how much we can change the constraints without changing the shadow prices.**

YOU TRY IT 3.5
Experiment with what happens when the constraints are changed past the limits in the right-hand-side ranging table. Try the labor constraint and the constraint on how many of the type two necklaces we should make in the necklace problem.

We also have ranges for the objective coefficients. Between those limits the values of the variables remain the same, in the necklace problem 14 necklaces of the first type and 24 of the second. Knowing this we can calculate the new profit (which will change). In other words, we can tell without resolving the problem how much the profit for a necklace can change before the solution changes.

1) The profit for necklaces of type 1 can go from $166.67 to as large as we want without changing the numbers of necklaces to be made in the optimal solution. So the total profit will change, but the variable values won't.

2) We will keep making the same number of necklaces even if the profit for the type 2 necklaces went as high as $300.00 each. Again note that the total profit will change!

> **The coefficient ranges tell us how much we can change the numbers in the objective without changing the final values of the variables.**

Thinking back to the graphical solution you can see why changing the objective function does not change the feasible region. The solution will still be one of the original corners. The numbers in the contribution ranging table tell how far each of the objective function coefficients can change without moving from that original optimal corner point.

IMPORTANT: All the sensitivity results we have discussed apply only to **changing one number at a time.** Even with this restriction there is a bountiful supply of information to be gathered in the output of a linear programming program above and beyond the optimal solution.

YOU TRY IT 3.6

For the necklace problem make changes in the objective function to experiment with how the solution changes. Make the profit for necklace 1 less than $166.67, and also very large. Make the profit on necklace 2 almost $300 and then more than $300. Record your results below.

Example 3.2: Suppose that Swales Jewelers (see Example 3.1) can get extra diamonds at $50 each for the two necklaces made.

a) How does the solution change? How many extra diamonds are purchased?

b) What is the unit worth of labor in this case?

c) If the Jewelers could sell left over emeralds to their customers for $60.00 each, how much would this add to the profit?

d) Does the answer change if the price of additional diamonds is $77.00 instead of $50.00?

Solution: Part of this problem is set up the same as in Example 3.1. However, we can purchase more diamonds which changes the problem significantly. Since how many diamonds we will purchase is unknown let's make it another variable.

```
N1 = the number of type one necklaces to make.
N2 = the number of type two necklaces to make.
 D = the number of diamonds to buy.
```

The fact that we can buy diamonds will affect the objective, because for each diamond we buy we pay $50.00, decreasing our profit. So we need the term $-50D$ in the objective.

```
maximize 200N1 + 250N2 - 50D
```

The chart for the constraints (where * means we need to make a change) is

	N1	N2	D	limit
Diamonds	2	3	*	100*
Emeralds	3	2	0	120
Labor	1	2		70
N2 Demand		1		24

Let's look at the diamonds carefully. The constraint is complicated by the fact that we can buy more diamonds "at 50 dollars each". So the right-hand-side of the constraint should be 100 plus the number of diamonds we buy. We can write this as: $2N1 + 3N2 \leq 100 + D$. Let's put this into standard form by subtracting D from both sides:

$$2N1 + 3N2 - D \leq 100$$

This gives the following math formulation:

```
Let N1 = the number of type one necklaces to make
    N2 = the number of type two necklaces to make
     D = the number of diamonds to buy

    maximize       Z = 200N1 + 250N2 - 50D
subject to diamonds:    2N1 +   3N2 -    D ≤ 100
           emeralds :   3N1 +   2N2      ≤ 120
           labor:        N1 +   2N2      ≤ 70
           N2 demand:           N2       ≥ 24
                   D ≥ 0   N1 ≥ 0   N2 ≥ 0
```

There are three variables, so we must use the software to produce a solution. Here is the solution for part a:

$$
\begin{aligned}
Z^* &= 9,600 \\
N1 &= 22 \\
N2 &= 24 \\
D &= 16
\end{aligned}
$$

So we should make 22 of the first type necklace, 24 of the second and purchase 16 additional diamonds. Profit will be $9,600.

To answer part b, we must consider the shadow prices:

Constraint	Shadow Price
Diamond	50.0
Emerald	0.0
Labor	100.0
N2 Demand	-100.0

From these numbers we see that the diamonds are worth $50 each (why is this not a surprise?), the emeralds are worth nothing and the labor is worth $100 for each hour. Thus the answer to part b is $100.00.

To answer part c, we need to consider the slack variables:

$$
\begin{aligned}
S1 &= 0.0 \\
S2 &= 6.0 \\
S3 &= 0.0 \\
S4 &= 0.0
\end{aligned}
$$

Since the second constraint concerned the emeralds, S2 = 6 says we have 6 left over, and so can sell them for a total of 60(6) = $360.00 added profit.

To answer part d, we must consider the sensitivity ranges for the objective coefficients:

```
          objective coefficient ranges
  variable          lower              upper
    N1              150.0           + infinity
    N2            - infinity          350.0
    D               -100               0
```

Given these ranges we can say that the number of necklaces made and the number of extra diamonds won't change as long as D stays between -100 and 0. So the answer doesn't change if the price of extra diamonds changes to $77.00, since -77 is between -100 and 0.

There are several things to note in this example. First, we have finally solved a problem with more than two variables in it. We couldn't have done this problem with our graphical approach of Section 2. The computer has no trouble, however, and can solve even larger problems.

Second, the "what if" questions we answered using the sensitivity information could have been answered by changing the numbers in the original problem and re-running the computer program. This approach might actually be faster until we're better at interpreting sensitivity information. What is the advantage of having the sensitivity information? Consider the following scenario:

A major airline wants to optimize their assignment of crews (pilots and cabin attendants) to routes, while keeping to the following guidelines:
- Crews should stay together.
- Crews should stop at least every third day at their home base.
- Maximum flight time in one day is limited; maximum flight time in one month is limited.
- The total weight limit on a crew must not exceed a certain limit.
- Certain pilots are trained on certain aircraft.
- Only a certain number of overseas flights are allowed per crew per month.

This problem turns out to be a huge linear programming problem. The variables are crew-routes (i.e., what crews are assigned to what routes) and the constraints come from the guidelines. If an airline has 500 routes and 1000 crews (not unreasonable for a large airline) the number of variables could be as large as 500,000! Even the fastest supercomputer is going to crank on this problem for a very long time. We certainly don't want to re-run the problem to answer every "what if" question the airline might have. Understanding the sensitivity information allows us to avoid most "re-runs."

YOU TRY IT 3.7

Use the information given about the solution of this problem to answer the questions.

A textile plant processes 2 types of cloth, acrylic and wool. Acrylic is available in unlimited supply, but only 1,000 thousand pound lots of wool can be stored at the plant per year. The weaving department requires 2 hours of production time per 1000 pounds of wool and 3 hours per 1000 pounds of acrylic. This department has 12000 available hours per year. The spinning room has 14100 available hours per year. The wool requires 4 hours per 1000 pounds in the spinning room and the acrylic requires 3 hours. The profit from the wool is $62 per 1000 pounds and $48 per 1000 pounds of the acrylic.

```
A = number of 1000s of pounds of acrylic made in a year
W = number of 1000s of pounds of wool made in a year

        maximize        62W + 48A
        subject to       2W +  3A  ≤  12000   weaving
                         4W +  3A  ≤  14100   spinning
                          W        ≤   1000   wool
                        W ≥ 0  A ≥ 0
```

Solution: Z* = 222000.00, W* = 1000.00, A* = 3333.33

Slacks: S(1) = 0.00, S(2) = 100.00, S(3) = 0.00

Shadow Prices: weaving 16.0, spinning 0.0, wool 30.0

```
           objective coefficient ranges
     variable          lower           upper
        W              32.00         + infinity
        A             - 0.00           93.00
```

a) How much of each type of fabric should be produced in a year to maximize profits?

b) The yarn shop across the street has offered to buy any extra time in the spinning room for $20.00 per hour. How much would the textile plant earn this way? Why?

c) Will the original answer change if the profit from the wool profit was lowered to $35 per 1000 pounds? Why?

d) The plant has some capital to expand the fabric making business. How should it spend the money: Adding more room in the spinning room, adding more space in the weaving room or adding more space in the wool storage facility? Why?

When interpreting the slacks and shadow prices for a problem, we must keep in mind whether the objective is to minimize or maximize and which kinds of inequalities we're dealing with, $\leq$ or $\geq$. We've done several examples where we've maximized with $\leq$ constraints, now let's examine what happens with the following minimization problem. The sensitivity ranges are interpreted the same way in both cases, so we'll concentrate on slacks and shadow prices.

Example 3.3: The Watauga county refuse department runs two recycling centers. In a typical day 140 pounds of glass, 75 pounds of newsprint, and 60 pounds of aluminum are deposited at Center 1, while 100 pounds of glass, 60 pounds of newsprint, and 180 pounds of aluminum are deposited per day at Center 2. The county has committed to deliver at least 1540 pounds of glass, 700 pounds of newsprint, and 1440 pounds of aluminum per week to encourage a recycler to open up a plant in town. Center 1 cost $40 and Center 2 $50 per day to operate, paying for one person to check on them both once an hour.

a) How many days per week should the county open each center to minimize it's cost and still meet the recycler's needs?

b) How are the slacks interpreted in this case?

c) How are the shadow prices interpreted in this case?

Solution: The linear programming translation is as follows.

```
c1 = number of days to open center 1
c2 = number of days to open center 2
```

Then we are asked to

```
minimize    40 c1 +  50 c2 (cost)
subject to
            140 c1 + 100 c2 ≥ 1540 (glass)
             75 c1 +  60 c2 ≥ 700  (newsprint)
             60 c1 + 180 c2 ≥ 1440 (aluminum)
```

The computer output contains the following information:

Solution: Z* = 505.00, c1* = 6.94, c2* = 5.69

Slacks: S(1) = 0.00, S(2) = 161.56, S(3) = 0.00

Shadow Prices:
       glass -0.25, newsprint 0.0, aluminum -0.08

a) The solution is to open Center 1 for 6.94 days per week and Center 2 for 5.69 days per week, giving a total cost of $505.00.

b) From the slack variables, we see that we will overproduce newsprint by 161.56 pounds above the recycler's needs, producing exactly the needed amounts of the other materials. The slacks give leftover amounts for $\geq$ constraints. For this reason they are sometimes called **surplus** variables in this situation.

c) Two of these shadow prices are negative. Ignore the negative sign for shadow prices; it is just there to remind us that we lose rather than gain in a minimize situation. Let's look at the shadow prices for this problem:

The shadow price of glass is -0.25. This says that each additional pound of glass the recycler requires the county to collect will **cost** an additional $0.25. We add 0.25 to the objective if the glass requirement is raised to 1541, and we subtract 0.25 from the objective (lowering the cost) if the glass requirement is lowered to 1539. Common sense tells us it should work this way!

The shadow price for newsprint is 0.0. As usual, this means that newsprint is not "worth" anything. We already collect more than the recycler needs, so if they ask for less we have it with no additional cost or savings.

The shadow price for aluminum is -0.08. As with glass, each additional pound of aluminum we need to collect for the recycler costs us $0.08 more.

YOU TRY IT 3.8

Use the computer generated solution to complete the following:

A farmer is custom mixing several store brands of fertilizer in order to come up with one that meets new guidelines on nitrogen-potassium-phosphorus requirements for his crops at a minimum cost. The three store brands he can buy are 12-8-6 (12 parts nitrogen, 8 parts potassium and 6 parts phosphorus), 10-10-10, and 20-2-2. These fertilizers cost $2.89, $2.95 and $4.62, respectively, per pound. He needs a fertilizer of at least 16-3-6 and no more than 6 parts potassium.

a) Suggest a blend that minimizes his cost.

b) Will he be imparting any extra nutrients to his crop?

c) Discuss the meaning of each shadow price in this situation.

Exercise Set 4.3

Solve the following problems using a linear programming package.

1) (Problem 12 from Section 1)   The State Employee's Credit Union (SECU) has $15.5 million in funds available for the mortgages and auto loans applied for by its members. The primary stockholders have recently approved the following policies regarding lending:

   • Keep at least 5 times as much money invested in mortgages as cars
   • Keep 18% of existing capital in short-term investments (i.e., not in loans).
   • For the next quarter, set interest rates at 10.25% for car loans and 8.25% for mortgages.

   Current loan applications include $4.2 million in car loans and $10.5 million in mortgages. How much should be allocated to each type of loan for the SECU to maximize its earnings?

2) (Problem 20 from Section 1)   Danbury Mining Corporation (DMC) owns two mines in Pennsylvania. The coal from the mines is separated into two grades before shipping. The Pittston mine produces an average of 10 tons of low-grade and 15 tons of high-grade coal daily at an operating cost of $10,000. The Altoona mine produces an average of 12 tons of low-grade and 8 tons of high-grade coal daily at a cost of $6200. Contractual obligations require DMC to ship 500 tons of low-grade and 600 tons of high-grade coal each month. Determine the number of days each mine must operate to meet the company's contracts at a minimum cost.

3) (Problem 13 from Section 1)   A nut company sells three mixes, Regular, Deluxe and Supreme, each with a $4 profit. Regular has 50% peanuts, 30% cashews and 20% hazelnuts. Deluxe has 30% peanuts, 40% cashews and 30% hazelnuts. Supreme has 20% peanuts, 40% cashews and 40% hazelnuts. If the company has at most 800 pounds of peanuts, 400 pounds of cashews and 295 pounds of hazelnuts available in a day, how many pounds of each mix should it make?

4) (Problem 23 from Section 1)   Metalheads, Inc., wishes to start production and sale of a new metal alloy containing iron, copper, and lead. The alloys containing iron, copper, and lead available to buy are A, B, C, D, and E. The percent of iron, copper, and lead in each and in the desired alloy are:

	A	B	C	D	E	needed
% iron	10	10	40	60	30	exactly 30
% copper	10	30	50	30	40	exactly 30
% lead	80	60	10	10	30	exactly 40
cost/lb	3.9	4.1	6.0	6.03	7.7	

   What is the best combination of the alloys A, B, C, D, and E to give exactly the needed percentages that will minimize the cost?

5) (Problem 22 from Section 1)   Two warehouses have canned soup on hand and three stores require soup in stock. UFT, United Food Transport, has contracted to deliver the soup. Warehouse 1 has 125 cases on hand and Warehouse 2 has 175 cases. The stores which we will call A, B, and C, all have needs of 100 cases each. The costs per case for shipping between the warehouses and stores are given in the table below. Find the solution that minimizes the cost of shipping.

	store A	store B	store C
warehouse 1	$0.08	$0.10	$0.28
warehouse 2	$0.10	$0.18	$0.15

6) Use the information about the solution given to answer the questions.

Suppose we choose to add another doll to the Barbie and Ken Problem. Skipper (Barbie's little sister) needs 9 ounces of plastic, 3 ounces of nylon (her hair isn't quite so "big"!), 4 ounces of cardboard, and she will have a profit of $5.95. How does this change the picture?

```
max    6.50 K + 6.00 B + 5.95 S           (OBJECTIVE)
s.t.    14 K +   12 B +    9 S ≤ 100,000  (PLASTIC)
                  5 B +    3 S ≤  30,000  (NYLON)
         4 K +    4 B +    4 S ≤  35,000  (CARDBOARD)
          K ≥ 0 AND B ≥ 0 AND S ≥ 0

Solution: K=4250, B=0, S=4500, Z=54400
Slacks:   S1 = 0, S2 = 16500, S3 = 0
Shadow Prices:   S1 = 0.11, S2 = 0, S3 = 1.24
```

	objective coefficient ranges	
K	6.03	9.26
B	-∞	6.28
S	5.25	6.5

	rhs ranges	
plastic	78750	122500
nylon	13500	+∞
cardboard	28571.4	42857.1

a) What is your recommendation (write the solution in a sentence.)

b) If we could get more plastic or cardboard, which should we get?

c) Which doll's price is "most sensitive" to changes?

d) The toy company can sell leftover nylon to the Hanes mill down the road for $.03 cents per ounce. How much additional cash can they make this way in a week?

7) Use the information about the solution given to answer the questions.

Melissa wants to maximize her income by investing the $120,000 her uncle left her as profitably as she can. She wants to invest part in a stock fund that yields 12% on average, and part in a bond fund that yields 7%. She is adverse to taking risks, and the stock market is a bit risky, so she wants no more money in stocks than in bonds. There is a 2% charge for purchasing the stock fund and a 1% fee for purchasing the bond fund. Melissa doesn't want to spend more than $1700 on these charges. She also wants to diversify by having at least $40,000 in each fund.

```
maximize .12 stock + .07 bond        (interest)
st.        stock + bond ≤ 120000     (total $$)
           stock - bond ≤ 0          (risk)
      0.02 stock + 0.01 bond ≤ 1700  (fees)
           stock ≥ 40000             (min in stock)
           bond ≥ 40000              (min in bond)

  solution:  Z* = 10900.0, stock = 50000.0, bond = 70000.0
```

	objective coefficient ranges	
stock	0.07	0.14
bond	0.06	0.12

a) How should Melissa invest the money?

b) Comment on the sensitivity of the interest rates: If the 12% and 7% rates are just rough estimates that could be off by as much as 1% either way, should Melissa be worried?

8)   (Problem 21 from Section 1)   A bag of dog food manufactured by Fido, Inc., must meet the guaranteed analysis printed on the bag: At least 14% crude protein, not less than 8% crude fat, at most 5.5% crude fiber and no more than 12% moisture. There are three ingredients that can be mixed to produce the tasty dog food that dogs prefer: I (yellow corn, bran) costs $0.45 per pound and contains 12% crude protein, 2% crude fat, 11% crude fiber and 20% moisture. II (chicken, beef tallow) costs $1.07 per pound and contains 26% crude protein, 31% crude fat, 1% crude fiber, and 8% moisture. III (ground wheat and brewer's rice) costs $0.60 per pound and contains 10% crude protein, 0% crude fat, 18% crude fiber and 6% moisture. What mix of the three ingredients -- I, II, and III -- should be mixed to total one pound for a minimum cost?

a) How should the mix be put together?

b) What is the cost of the best mixture?

c) Which nutritional requirements did we satisfy exactly and which did we go over/under?

d) The veterinary association released new guidelines on dog nutrition, and the only change was that the crude fiber should now be at most 4.5%. What can we tell the company about how the answer will change?

e) The veterinary association released different new guidelines on dog nutrition, and the only change was that the crude protein should now be at least 25%. What can we tell the company about how the answer will change?

f) The price on ingredient III sky-rocketed to $2.00 per pound. How will this affect the cost of the mixture? Is the old mixture still the best?

9)   (Problem 15 from Section 1)   The Coswell Coat Company (CCC) makes 3 kinds of jackets: Cotton-filled (CF), wool-filled (WF) and goose-down filled (GF). Each involves 50 minutes of sewing, but it takes 8 minutes to stuff a CF, 9 minutes to stuff a WF and 6 minutes to stuff a GF. The factory has a supplier that will furnish it 225 pounds of wool and 135 pounds of goose-down and all the cotton that is needed per week. It takes 12 ounces total to fill a jacket. The factory has 666 and 2/3 hours of sewing time available and 80 hours of filling time available per week. It makes $24 on each CF jacket, $32 on each WF and $36 on each GF. How many of each should be made to maximize profit?

a) What is the solution?

b) If the company could hire one more person, which would be more profitable, a sewer or a stuffer? Why?

c) If the price of a cotton-filled jacket is raised so that profit becomes $30, will the number of each type of coat made change? If so, to what?

d) Due to a drought, the cotton supply becomes limited to 260 pounds per week. How does this change the solution? Can you tell without resolving the problem?

10)   (Example 2.4 from Section 2)   A chemical plant needs to cut its emission of sulphur dioxide (SO_2) and suspended particulate matter to satisfy federal

regulations. It can use scrubbers in the smokestack or filters in the process. A filter can reduce particulate matter by 12 pounds per hour and SO_2 emissions by 2 pounds per hour. A scrubber reduces particulate matter by 7 pounds per hour and SO_2 emissions by 4 pounds per hour. Filter units cost $4000 and scrubber units cost $6000. The smokestack can hold only 12 scrubbers and no more than 10 filter units can be installed. How can the plant reduce SO_2 emissions by at least 44 pounds per hour and particulate emissions by at least 84 pounds per hour at least cost?

The linear programming translation is as follows.

```
F = the number of filters to install
S = the number of scrubbers to install
minimize 4000F + 6000S
subject to
        (1) SO2              2F + 4S ≥ 44
        (2) Particulate     12F + 7S ≥ 84
        (3) Max scrubbers         S ≤ 12
        (4) Max Filters      F        ≤ 10
```

The sensitivity information is as follows.

```
solution:  Z* = 66834.0, F = .8235, S = 10.59

slacks:S(1) = 0, S(2) = 0, S(3) = 1.41, S(4) = 9.18

shadow prices: SO2: -1294.12, particulate: -117.65, Max
scrubbers: 0, Max filters: 0
```

```
         objective coefficient ranges
    F   3000        10,285.71
    S  2333.33         8000
```

```
         right-hand-side ranges
        SO2          14.0       48.0
    particulate      77.0      162.0
    max scrubber    10.58     infinity
     max filter      .8235    infinity
```

a) What would be the cost of having to cut the amount of SO_2 by 1 more pound per hour?

b) What would be the cost of having to cut the amount of particulate by one more pound per hour?

c) How much would it be worth to have space available for another scrubber?

d) Over what range will the shadow price of removing SO_2 remain constant?

e) What is the minimum amount of particulate that we would have to remove and still not have the shadow price for the particulate constraint change?

f) How high would the cost of scrubbers have to go before the number of each to install would change?

g) What would you tell someone that the cost of removing sulfur dioxide is? The cost of removing particulate matter? Should you look at the $66,834 or at the shadow prices?

11) (Problem 16 from Section 1) A manufacturing firm wants to consider making 3 new styles of recliners: the deluxe, the easyrider and the economy model. The times left over from other products in the various shops are given:

shop	available time (hours per week)
woodwork	500
finishing	500
upholstery	300

The number of machine hours needed to make one of each type of recliner are:

shop	deluxe	easyrider	economy
woodwork	10	4	5
finishing	4	5	0
upholstery	1	0	3

The sales department did a market survey and found that the sales potential for the deluxe and easyrider is more than the company can make in a week, but the maximum sales potential for the economy is 25 per week. The profit for each is:

chair:	deluxe	easyrider	economy
profit:	$60	$50	$25

a) What is the solution?

b) What is time in each shop worth? If hours could be added in one shop, which should it be? Why?

c) If available time in the finishing shop is increased to 525 hours per week, does the finishing time become less valuable? Why?

d) How much can the price (and hence the profit) of the economy model be reduced by without changing the number of each type of chair made?

12) Boone paint mixes custom colors from 3 base pigments by adding them to white paint. The base pigments are: red, yellow and blue. They have a total of 600 ounces of red pigment, 300 ounces of yellow pigment and 200 ounces of blue pigment available in any given week. Each gallon of custom color costs the store $6.00 to make. The custom colors that they make are:

1) green - sells for $10.00 per gallon - requires 3 oz blue and 4 oz yellow
2) purple - sells for $11.00 per gallon - requires 5 oz red and 2 oz blue
3) pink - sells for $9.00 per gallon - requires 3 oz red only
4) orange - sells for $8.00 per gallon - requires 4 oz yellow and 1 oz red

a) How many gallons of the custom colors should they make in one week to maximize profit given the pigment constraints? (you may assume that they can sell all they make!)

b) If the price of the orange paint is increased by $2.00 per gallon, will the amount of custom paints made change?

c) If the store increases its supply of one of the pigments, which would be most cost effective? Why?

d) Boone Art Supply will buy up any left over pigments that Boone Paint can spare. If they agree to buy leftovers at 40 cents per ounce, how much will Boone Paint earn this way?

13) The Boone Brewery makes four products called light, dark, ale and premium. These products are made using water, malt, hops, and yeast. The brewery has a free supply of water but the amount of the other resources restrict the production. The table below gives the pounds of each resource required to produce one vat of each product, the pounds of each resource available each week, and the profit received for each vat of product. How much of each product should be made to maximize profit?

	light	dark	ale	premium	available per week
malt	1	1	0	3	50 lb
hops	2	1	2	1	150 lb
yeast	1	1	1	4	80 lb
profit	$6	$5	$3	$7	

a) What is the solution?

b) How much must the profit of the premium be increased to make it worth producing?

c) If we are limited (due to a fire in the warehouse) to storing no more than 100 pounds of hops, how will the solution change?

d) Which raw material (malt, hops, yeast) should we increase the supply of to increase profit the most? Why?

14) A manufacturer of bicycles builds one-speed, three-speed, and ten-speed models. The bicycles need both aluminum and steel. The company has available 91,800 units of steel and 42,000 units of aluminum. The one-, three-, and ten-speed models need, respectively, 20, 30, and 40 units of steel and 12, 21, and 16 units of aluminum. How many of each type of bicycle should be made in order to maximize profit if the company makes $8 per one-speed bike, $12 per three-speed bike, and $24 per ten-speed bike?

a) Find the solution.

b) If the profit from the three-speed bike is reduced to $8, does the number of bikes made change?

c) How much the company should be willing to pay for extra steel and aluminum?

d) What is the absolute most steel that it would make sense for the company to buy if the other conditions don't change? Use the sensitivity output and some experiments to decide.

15) A furniture company manufactures 4 kinds of desks. Each desk is first constructed in the carpentry shop and is then sent to the finishing shop where it is varnished, waxed and polished. The number of man hours of labor required is as follows:

desk 1	4 hours carpentry	1 hour finishing
desk 2	9 hours carpentry	1 hour finishing
desk 3	7 hours carpentry	3 hour finishing
desk 4	10 hours carpentry	40 hour finishing

Because of limitations in capacity of the plant, no more than 6000 man hours can be expected in the carpentry shop and 4000 in the finishing shop in the next 6 months. The profit from the sale of each desk is: $12 for desk 1, $20 for desk 2, $18 for desk 3, $40 for desk 4. Assuming that raw materials and supplies are available in adequate supply and all desks produced are sold, the company wants to determine how many of each type of desk to make to maximize the profit.

a) Translate and solve this problem.

b) As manager you are given permission to hire more workers. What area should you put them in supposing you could hire trained workers in any area?

c) What profit must be made from desk 2 to make it worth producing?

d) If the company decides to phase out production of desk 1, will the solution change? What would the new solution be? Can you find this without resolving the problem?

---

4   INTEGER PROGRAMMING

In all of the problems so far, we have assumed that the **divisibility** property holds: Fractional answers are always allowed. In several we explained fractional answers as averages. For example, suppose we solve a problem about manufacturing three kinds of tables and arrive at the optimal solution:

Make 100.2 of table 1, 87 of table 2 and 55 of table 3 each week.

We can get around the divisibility requirement by saying that we make *on average* 100.2 of table 1. What does this imply? Something like this would work:

Week 1: Make 100
Week 2: Make 100
Week 3: Make 100
Week 4: Make 100
Week 5: Make 101
Average: 100.2

What if our problem deals with a one time occurrence? Then this rationalization of fractions is not appropriate. What can we do? As we will see, rounding is not always the best approach. Consider the following problem:

$$\text{maximize} \quad Z = x + 5y$$
$$\text{subject to:}$$
$$x + 10y \leq 20$$
$$x \leq 2$$
$$x \geq 0, \; y \geq 0$$

The graphical solution to this is easy:

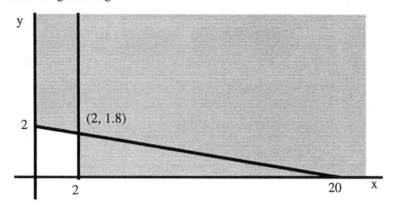

The objective evaluated at the corners is:

x	y	obj
0	0	0
2	0	2
0	2	10
2	1.8	11

From this we would state the solution as $x = 2$, $y = 1.8$, $Z = 11$. Now let's look at this again assuming that we must have an integer solution. Rounding 1.8 up and down gives:

$x = 2$, $y = 2$ oops, this is outside the region!
$x = 2$, $y = 1$ $Z = 7$

Hum. This would lead us to believe that $x = 2$, $y = 1$ and $Z = 7$. Let's look at the feasible region again, this time only plotting the integer pairs that satisfy the constraints (black dots are in the region; white dots are not):

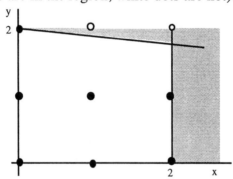

So all the feasible integer points are:

x	y	obj
0	0	0
1	0	1
2	0	2
0	1	5
1	1	6
2	1	7
0	2	10

Clearly, the optimum is $x = 0$, $y = 2$ and $Z = 10$. Interesting! Even though the x variable had an integer value of 1 in the regular solution, in the integer solution the value of x changes to 0. Try one yourself:

YOU TRY IT 4.1

Solve the problem given below as a linear programming problem and then as an integer programming problem. Are the answers the same?

```
maximize    Z = 8 x + 5 y
subject to:
            x +   y ≤ 6
            9 x + 5 y ≤ 45
            x ≥ 0, y ≥ 0
```

What technique can we use to find integer answers if rounding isn't going to work? Is there a method like the simplex method that will hop onto an edge and walk to the best point? No. However, we can use educated guessing and linear programming to help us in an approach called the **Branch and Bound Method**.

Here are a few facts about linear programming that will come in handy. None of these are difficult to justify. Try to explain them!

- The feasible region for an Integer Programming Problem is always contained in the feasible region for the corresponding Linear Programming Problem.

- The maximum for an Integer Programming Problem will always be less than or equal to the maximum for the corresponding Linear Programming Problem.

- If a constraint gets added, the maximum for the new problem cannot be larger than the maximum for the original problem.

Here's the idea behind branch and bound:

1. Solve the corresponding LP problem (by hand or on the computer). If the variables that need to be integer are, stop.
2. If a variable that needs to be integer is fractional, "branch" on that variable by creating 2 new sub-problems: If the variable is x and the value is A:
   - create sub-problem 1 by adding the constraint x ≤ (round A down)
   - create sub-problem 2 by adding the constraint x ≥ (round A up)
3. Solve these new problems; if integer answers come up, choose the best. If a variable which should be integer is not, branch again on that variable.
4. Repeat this process until the best integer answer is found.

That is a lot to digest. Let's try this with our original example. The problem was:

```
maximize     Z = x + 5 y
subject to:
             x + 10 y ≤ 20
             x ≤ 2
             x ≥ 0, y ≥ 0
```

The solution was: x = 2, y = 1.8, Z = 11. Y needs to be integer, so we need to branch on y; consider the two sub-problems:

```
Sub-problem 1                    Sub-problem 2
maximize    Z = x + 5y           maximize     Z = x + 5y
subject to:                      subject to:
            x + 10y ≤ 20                      x + 10y ≤ 20
            x ≤ 2                              x ≤ 2
            y ≤ 1  (1.8 rounded down)          y ≥ 2  (1.8 rounded up)
            x ≥ 0, y ≥ 0                       x ≥ 0, y ≥ 0
```

Notice the new constraints in bold type. The graphical solutions to these problems (try them and see; the second problem is very interesting because the feasible region is one point!) are:

```
Sub-problem 1                    Sub-problem 2
x = 1, y = 1, Z = 5              x = 0, y = 2, Z = 10
```

We have two integer answers to pick from and the second one has the better (higher) objective value so it must be the solution. Why can we say this? What did branching on the variable y do to the original problem? It subdivided the original feasible region into two pieces cutting out a section that contains only fractional y values:

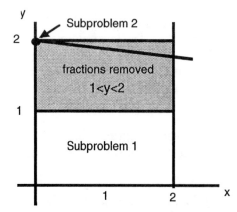

So what did we do? We removed a strip from the feasible region. This strip contains fractional y values only ($1 \leq y \leq 2$). That breaks the original region into two pieces, one of which must contain the answer. These regions correspond to the two sub-problems, so of the sub-problems must contain the optimal answer for the original problem.

Here are a couple more for you to try. Notice that we only require one of the variables to be integer in the first one. Pay careful attention to the hint on the second one. It has eight variables, so you'll have to have a computer handy to solve the linear programming problems.

YOU TRY IT 4.2
Solve the problem given below using branch and bound.

```
maximize    Z = 2p + q
subject to:
            5p + 2q ≤ 8
            4p + 4q ≤ 12
            p ≥ 0, q ≥ 0
            p integer
```

YOU TRY IT 4.3
Gotham City Bank is open Monday through Friday from 9 to 5. From past experience the bank manager knows that during a given hour she needs at least the number of tellers given below.

Time:	9	10	11	12	1	2	3	4
Tellers:	4	3	4	6	5	6	8	8

The bank hires two kinds of tellers: Full time tellers work 9 to 5 five days a week, with one hour off for lunch. The bank determines when each teller goes to lunch, with some going at noon and some at 1. Full time employees are paid $8 per hour. The bank also is willing to hire up to five part time tellers. Each part time teller works exactly 3 consecutive hours per day five days a week. A part time teller is paid $5 per hour. How many of each type teller should be hired to minimize cost? Hint: Let F1 = full timers who lunch at noon, F2 = full timers who lunch at 1, P1 = part timers who start at 9, P2 = part timers who start at 10, P3 = part timers who start at 11, P4 = part timers who start at 12, P5 = part timers who start at 1, P6 = part timers who start at 2. There's a constraint for who's working during each hour and for the number of part time tellers to be ≤ 5.

Unfortunately, it isn't always the case that one branch finds the solution. Here's one that takes more branching with the added twist of needing computer solution.

Example 4.1: Find the solution to the problem below assuming that x, y, and z must be integer but w can be anything.

```
maximize    Z = 4 x - 2 y + 7 z - w
subject to
            x + 5 z ≤ 10
            x + y - z ≤ 1
            6 x - 5 y ≤ 0
            - x + 2 z - 2 w ≤ 3
            x, y, z, w ≥ 0
```

Solution: The computer solution to the linear programming problem is:

```
    x = 1.25, y = 1.5, z = 1.75, w = 0, Z = 14.25
```

x, y, and z need to be integer, so we'll have to branch. What should we branch on? Just choose one of the non-conformists, say x:

Now we need to solve sub-problems 1 and 2:

```
    Sub-problem 1: add  x ≤ 1
    Sub-problem 2: add  x ≥ 2
```

The solutions (using the computer again) are:

```
    Sub-problem 1: x=1, y=1.2, z=1.8, w=0, Z=14.2
    Sub-problem 2: not feasible
```

What now? Sub-problem 2 is a dead end; we refer to that part as being *fathomed*. Sub-problem 1 still doesn't have integers where we need them. So we need to branch again, this time using y or z; let's use y:

```
    Sub-problem 3: add  x ≤ 1 and y ≤ 1
    Sub-problem 4: add  x ≤ 1 and y ≥ 2
```

The solutions (using the computer again) are:

```
    Sub-problem 3: x=0.8333, y=1, z=1.833, w=0, Z=14.17
    Sub-problem 4: x=0.8333, y=2, z=1.833, w=0, Z=12.17
```

Hum. We're still not done, and we can't throw out either of these. Let's work with sub-problem 3: We continue branching this time on x again to get sub-problems 5 and 6.

```
    Sub-problem 5: add  x ≤ 1 and y ≤ 1 and x ≤ 0
    Sub-problem 6: add  x ≤ 1 and y ≤ 1 and x ≥ 1
```

The solutions (using the computer again) are:

```
    Sub-problem 5: x=0, y=0, z=2, w=0.5, Z=13.5
    Sub-problem 6: not feasible
```

Here is a tree-diagram to illustrate the steps so far:

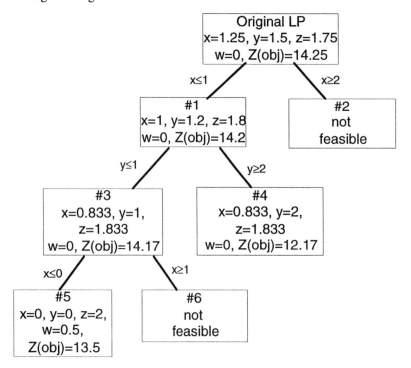

At this point we have found a candidate for the answer with sub-problem 5. Notice that x, y, and z are integer (and w doesn't have to be). To be sure we have found the best integer answer we have to go back and look at sub-problem 4. Here's where those facts from the beginning of the section will come in handy. One of them said that the integer answer in a maximize problem is always less than or equal to the corresponding LP problem's solution. Well, look at sub-problem 4. Its objective value is 12.17. The objective value for any integer branches would have to be no bigger than 12.17, so can any branches from here possibly win? No, since the objective for sub-problem 5 is 13.5.

We refer to sub-problem #4's branch as being fathomed. This is where the *bound* part of the name Branch and Bound comes from. So sub-problem 5 is the optimal integer solution.

The integer answer is: x = 0, y = 0, z = 0, w = 0.5, Z(obj) = 13.5.

We made two arbitrary choices in the solution above: Branching on x instead of y or z, and working with sub-problem 3 rather than 4 first. Choosing differently would not have changed the final solution, but it could take more or fewer branches if we chose differently. Try it and see.

Can we make these branching decisions in a smart way? Sometimes remembering what the variables represent can help in the decision on which variable to choose. If there is no obvious choice, just choose any variable which needs to be integer and isn't.

There *is* a strategy for choosing which sub-problem to work on that can save steps sometimes: Choose to work on the sub-problem with the higher objective value. Then work in the tree diagram all the way down one branch until finding a solution candidate. That way we might be able to eliminate some branching when we work our way back up the tree, taking care of branchings we skipped.

Before ending our discussion of integer programming problems, we should ask whether problems like the teller problem are just flukes. Are there so few situations like this that in general we can avoid the branch and bound approach? Here are a few examples of special problems called **decision problems** which all can be formulated as integer problems. After seeing a few you will be able to think of many situations like these.

Example 4.2: Consider the following puzzle: You are to pick out 4 of the three-letter "words" from the following list:

DBA   DEG   ADI   FFD   GHI   BCD   FDF   BAI

For each word you earn a score equal to the position in which the word's third letter occurs in the standard alphabet (A = 1, B = 2, etc.). Your goal is to choose the four words that maximize your total score subject to the following constraint: The sum of the positions in the alphabet for the first letters in the words chosen must be at least as large as the sum of the positions of the second letters in the chosen words.

Solution: We need to decide which of the words to choose to win. An easy way to formulate a yes-no decision like this is to use a yes-no variable. Let $x_1, x_2, x_3, x_4, x_5, x_6, x_7$, and x_8 be the variables, one for each of the words, and let

$$x_i = \begin{cases} 1 & \text{if word i is chosen} \\ 0 & \text{otherwise} \end{cases}$$

Now, let's formulate the objective and constraints:

The objective terms should be each variable times its third letter position. For the first word A is the third letter and A's position is 1; for the second word G is the third letter and G's position is 7, etc:

$$Z = 1x_1 + 7x_2 + 9x_3 + 4x_4 + 9x_5 + 4x_6 + 6x_7 + 9x_8$$

We need to choose exactly four words:

$$x_1 + x_2 + x_3 + x_4 + x_5 + x_6 + x_7 + x_8 = 4$$

We need to have the first letter positions add up to at least the total of the second letter positions:

$$4x_1 + 4x_2 + 1x_3 + 6x_4 + 7x_5 + 2x_6 + 6x_7 + 2x_8$$
$$\geq 2x_1 + 5x_2 + 4x_3 + 6x_4 + 8x_5 + 3x_6 + 4x_7 + 1x_8$$

Last we need to note that all of our variables are ≥ 0 and ≤ 1.

This type of yes-no or 0-1 variable is referred to as a **binary** variable. If all of the variables in a problem are binary, we call the problem a **Binary Integer Programming Problem** (BIP). Here's another example where a binary variable can help in a decision problem.

Example 4.3: The Boone Manufacturing Company is considering expansion by building new factories in either Raleigh or Charlotte, or both. It is also considering building a new warehouse in a city where a factory is built, or not building any warehouses. Here are the data collected by the company on costs and profitability for the choices:

	Variable	Profitability	Cost to Build
Factory in R	FR	$8 million	$5 million
Factory in C	FC	$10 million	$7 million
Warehouse in R	WR	$6 million	$2 million
Warehouse in C	WC	$9 million	$8 million

If the company has $10 million dollars in capitol for building, what should it do?

Solution: Each variable is a yes-no variable as in the previous example. What is the objective? It is not stated, so we will assume that the company wants to maximize profitability:

$$Z = 8\,FR + 10\,FC + 6\,WR + 9\,WC$$

Let's do the easiest constraint first: There is at most $10 million to spend:

$$5\,FR + 7\,FC + 2\,WR + 8\,WC \le 10$$

Now, what's left?

We will build either one or both of the factories:

$$1 \le FR + FC \le 2$$

We will build at most one warehouse:

$$WR + WC \le 1$$

If we build a warehouse in a city then we must build a factory there, too. So we need an inequality that forces FR to be 1 if WR is, but they are not necessarily equal. Check these:

$$FR \ge WR$$
$$FC \ge WC$$

In the first one, if WR is 1 then FR would have to be 1 as well. If WR was 0 then FR could be either 0 or 1. This is exactly what we want.

Last we need to note that all of the variables are ≥ 0 and ≤ 1.

The "If we build a warehouse in Charlotte, then we must have built a factory in Charlotte" constraint can be generalized to any **if-then statement**:

"If g is chosen then f must be as well"
can always be formulated as
$f \ge g$.

One other common situation that occurs is an **either - or but not both** problem. We handle this with an additional helper binary variable y:

"Either f or g but not both"
can always be formulated as
$f \le 1- y$ and $g \le y$.

If y is 0 then g is not chosen, and if y is 1 then f is not chosen. If we must choose at least 1, we could change the inequalities to equalities.

Let's try another one.

Example 4.4: Boone Custom Autoworks is considering manufacturing three types of cars: a compact, a sports car and a truck. The resources required are:

	compact	sporty	truck
steel	1.5 tons	3 tons	5 tons
labor	30 hours	25 hours	40 hours
profit	$2000	$3000	$4000

Currently 6000 tons of steel and 60,000 hours are available. For production of a car to make up for the cost of tooling the factory, at least 1000 of a type of car must be made. What should Boone Custom Autoworks do?

> <u>Solution:</u> On the surface, this seems like any old LP! Let's set up the objective and constraints using C, S and T as the number of compacts, sporties and trucks to make, respectively.
>
> The objective: `maximize Z = 2000C + 3000S + 4000T`
> The steel and labor constraints:
>
> ```
>      1.5C + 3S + 5T ≤ 6000
>      30C + 25S + 40T ≤ 60000
> ```
>
> Now we are at the "glitch." We need to turn off a variable if we make less than 1000 of that type car.
>
> Let's look at compact cars: We need to have either c ≥ 1000 or c ≤ 0. Consider this approach: Let IC be a binary variable, and M be a really large number:
>
> ```
>          C ≤ M IC
>      C ≥1000 - M ( 1 - IC )
> ```
>
> This seems weird but look at what happens:
>
> If IC is 1 then the first constraint says C is ≤ some really large number and the second constraint says that C is ≥ 1000 (+ 0).
>
> If IC is 0 then the first one says that C is ≤ 0 and the second one says that C is ≥ some really negative number.
>
> In summary, IC = 1 implies that C ≥ 1000 and IC = 0 implies that C = 0, just what we want to allow for.
>
> We do the same thing for S and T. We'll call their binary variables IS and IT. Here is the whole setup:
>
> ```
>      maximize Z = 2000C + 3000S + 4000M
>      subject to:
>              1.5C + 3S + 5M ≤ 6000
>              30C + 25S + 40M ≤ 60000
>                      C ≤ M IC
>              C ≥1000 - M ( 1 - IC )
>                      S ≤ M IC
>              S ≥1000 - M ( 1 - IS )
>                      T ≤ M IC
>              T ≥1000 - M ( 1 - IT )
>          C, S, T ≥ 0, IC, IS, IT binary
> ```

When we solve this problem on the computer, we'll need to specify a value for M; we can't just say "make M a large number." What value should we choose for M? It needs to be large enough so that it doesn't add a restriction we didn't intend. Let's look once more at how M affects a variable, say C:

$$C \leq M \; IC$$
$$C \geq 1000 - M \; (\; 1 \; - \; IC \;)$$

From the other constraints on labor and steel, we know that the total number of compact cars we could possibly make (if we make only compacts) is 2000, because any more and we'd run out of labor hours. So in the first inequality above, if M is 2000 we'd be safe. We can do the same analysis for the other two variables, and then either use different M's for each variable or just choose the largest M, which would work for all.

YOU TRY IT 4.4
Janet has $14,000 to invest and is considering four investments. One restriction on these investments is that if investment 2 is chosen then investment 3 must be chosen also. Find a strategy that will maximize her net present value. Here are the data:

	Initial Investment	Net Present Value
Investment 1	$5,000	$16,000
Investment 2	$7,500	$22,000
Investment 3	$4,000	$10,000
Investment 4	$2,500	$9,000

Integer Programming problems, binary programming problems, and mixed integer programming problems are so prevalent that most software packages which can solve linear programming problems have a built in automated branch and bound algorithm. Access to such a package makes solving problems in which integer answers are required even easier. Try some of the problems from the previous sections to see how the integer answers compare to the original answers. The homework set below has several problems like this.

Exercise Set 4.4

Find integer solutions to the following problems.

1) maximize $Z = 5 x + 2 y$
    subject to

$$3 x + y \leq 12$$
$$x + y \leq 5$$
$$x, y \geq 0, x, y \text{ both integer}$$

2) maximize $Z = 4 x + 3 y$
    subject to

$$4 x + 9 y \leq 26$$
$$8 x + 5 y \leq 17$$
$$x, y \geq 0, x, y \text{ both integer}$$

3) maximize $Z = 4 x + 3 y + z$
    subject to

$$3 x + 2 y + z \leq 7$$
$$2 x + y + 2 z \leq 11$$
$$x, y, z \geq 0, y, z \text{ both integer}$$

4) (Example 5.2) Consider the following puzzle: You are to pick out 4 of the three-letter "words" from the following list:

    DBA   DEG   ADI   FFD   GHI   BCD   FDF   BAI

    For each word you earn a score equal to the position in which the word's third letter occurs in the standard alphabet ($A = 1, B = 2$, etc.). Your goal is to choose the four words that maximize your total score subject to the following constraint: The sum of the positions in the alphabet for the first letters in the words chosen must be at least as large as the sum of the positions of the second letters in the chosen words.

5) (Example 5.3) The Boone Manufacturing Company is considering expansion by building new factories in either Raleigh or Charlotte, or both. It is also considering building a new warehouse in a city where a factory is built, or not building any warehouses. Here are the data collected by the company on costs and profitability for the choices:

	Variable	Profitability	Cost to Build
Factory in R	FR	$8 million	$5 million
Factory in C	FC	$10 million	$7 million
Warehouse in R	WR	$6 million	$2 million
Warehouse in C	WC	$9 million	$8 million

If the company has $10 million dollars in capitol for building, what should it do?

6) (Problem 8 Section 1) A manufacturer makes 2 kinds of cars, the sedan model and the sports model. According to a contract, it must supply at least 5000 cars. The sports model has a special engine of which only 2500 are available. Both models are made on an assembly line on which only 15,000 hours of time are available. The sedan takes 2 hours on the line and the sports car 4 hours. The profit is $750 on the sedan and $2000 on the sports model. How many of each should be made to maximize profit?

7) (Problem 15 Section 1) The Coswell Coat Company (CCC) makes 3 kinds of jackets: Cotton-filled (CF), wool-filled (WF) and goose-down filled (GF). Each involves 50 minutes of sewing, but it takes 8 minutes to stuff a CF, 9 minutes to stuff a WF and 6 minutes to stuff a GF. The factory has a supplier that will furnish it 225 pounds of wool and 135 pounds of goose-down and all the cotton that is needed per week. It takes 12 ounces total to fill a jacket. The factory has 666 and 2/3 hours of sewing time available and 80 hours of filling time available per week. It makes $24 on each CF jacket, $32 on each WF and $36 on each GF. How many of each should be made to maximize profit?

8) (Problem 17 Section 1) Mary, who is ill, takes mineral supplements. Each day she must have at least 21 units of iron, 10 units of potassium, and 40 units of calcium. She can choose between pill #1 which contains 7 units of iron, 2 of potassium, and 4 of calcium, and pill #2 which contains 3 units of iron, 2 of potassium, and 14 of calcium. Pill #1 costs $0.20 and pill #2 costs $0.35. Find the number of each pill which would minimize cost for Mary.

9) (Problem 6 Section 3) Suppose we choose to add another doll to the Barbie and Ken Problem. Skipper (Barbie's little sister) needs 9 ounces of plastic, 3 ounces of nylon (her hair isn't quite so "big"!), 4 ounces of cardboard, and she will have a profit of $5.95. How does this change the picture?

10) (Problem 12 Section 3) Boone paint mixes custom colors from 3 base pigments by adding them to white paint. The base pigments are: red, yellow and blue. They have a total of 600 ounces of red pigment, 300 ounces of yellow pigment and 200 ounces of blue pigment available in any given week. Each gallon of custom color costs the store $6.00 to make. The custom colors that they make are:

    1) green - sells for $10.00 per gallon - requires 3 oz blue and 4 oz yellow
    2) purple - sells for $11.00 per gallon - requires 5 oz red and 2 oz blue
    3) pink - sells for $9.00 per gallon - requires 3 oz red only
    4) orange - sells for $8.00 per gallon - requires 4 oz yellow and 1 oz red

How many gallons of the custom colors should they make in one week to maximize profit given the pigment constraints? (you may assume that they can sell all they make!)

11) Boone Ale Brewery is considering manufacturing three types of beer: a pale ale, a lemon ale and an amber. The ingredients required are (in pounds per vat) and profit per vat are:

	pale ale	lemon	amber
yeast	1	1	1
hops	2	1	3
malt	0	1	1
profit	100	80	110

60 pounds of malt, 145 pounds of hops and 85 pounds of yeast are available each week. For production of a beer to make up for the cost of refurbishing the equipment, at least 50 vats of it must be produced. What should the brewery do?

12) Mark has $25,000 to invest and is considering four investments. One restriction on these investments is that if investment 4 is chosen then investment 2 must be chosen also. Find a strategy that will maximize her net present value. Here are the data:

	Initial Investment	Net Present Value
Investment 1	$6,000	$16,000
Investment 2	$2,700	$4,200
Investment 3	$4,200	$10,000
Investment 4	$7,800	$22,000

13) An airline needs to assign its three crews to cover all of its regular flights. Exactly three of the possible flight sequences shown below need to be chosen (one for each crew) so that every individual flight is covered. The cost of assigning a crew to a flight sequence is also given. Formulate the problem as a decision problem and find the minimum cost.

	Possible Sequences											
Flight	1	2	3	4	5	6	7	8	9	10	11	12
NY - Phila	1			1			1			1		
NY - DC		1			1			1			1	
NY - Atlanta			1			1			1			1
Phila - Orlando				2			2		3	2		3
Phila - NY	2					3				5	5	
Orlando - DC				3	3				4			
Orlando - Atlanta							3	3		3	3	4
DC - NY		2		4	4				5			
DC - Orlando					2			2			2	
Atlanta - DC			2				4	4				5
Atlanta - Phila						2			2	4	4	2
Cost (in $1000s)	2	3	4	6	7	5	7	8	9	9	8	9

14) GasCo delivery trucks contain five compartments, holding up to 2700, 2800, 1100, 1800, and 3400 gallons of fuel, respectively. The company must deliver three types of fuel (ultra, super and unleaded) to a customer. The demands (in gallons), penalty per gallon short and maximum allowed shortage (in gallons) are:

type	demand	penalty per gallon short	maximum allowed shortage
ultra	2900	$10	500
super	4000	$8	500
regular	4900	$6	500

Each compartment of the truck can only carry one type of fuel. Find a filling schedule that minimizes shortage costs.

CHAPTER SUMMARY 5

We have seen how the algebra of linear inequalities is pervasive in business and industry decision making. Few choices are made without some form of mathematical analysis similar to the methods presented in this Chapter. Yet, with just a little practice, any person can analyze a situation, formulate the mathematical model and find a solution. One final word of warning: Don't forget your common sense! It is hard to code common sense into the math equations, so it's your job to decide how to interpret the answer for the given situation.

Key Ideas

Formulating the Mathematics

- Here are the steps one more time:

Translating a Linear Programming Problem
1. **Read** the problem carefully. Underline key words, then read it again.
2. Identify the **variables** and write down what they stand for. The Barbie and Ken problem asked "How many Barbies and Kens should they manufacture each week to maximize their profit?" That meant that the number of Barbies and the number of Kens were the variables. Remember that variables are things that can change, so if you have a **fixed** amount of something it can't be a variable. Name variables with letters that signify what they stand for, like B for the number of Barbies. Write out the units for the variables.
3. Find the **objective function**. Normally, there is a cost or profit associated with each unknown and the objective function is composed of sum of the value per item times the variable for that item, like 6.5K is the profit per Ken times the number of Kens. Is it a maximization or a minimization?
4. Find the **constraints**, things that restrict, or limit. If a process uses hours of labor, time on a machine, material, or anything else that is in short supply, it will generate a constraint. Summarize the constraint quantities in a table, giving each constraint a name. If there are "300 available" expect a constraint: something ≤ 300. If there must be "at least 200" expect: something ≥ 200. The left-hand-side will be a sum of numbers times variables, where the numbers are how much of the limited item it takes to produce one unit of the variable.
5. Go over the model and the words. **Be careful about units**. Each inequality must have consistent units between the terms on the left-hand-side and the right-hand-side. If hours and minutes both appear in the problem decide which to use and change everything to that. Carefully consider each constraint in words and see that it is expressed correctly as an inequality.

Solving Graphically

- Here are the steps one more time:

Solving a Linear Programming Problem using Graphing
1. Build a table of values for the line associated with each inequality constraint.
2. Draw the axes, using a scale that will work for all the constraints, and then graph each of the lines.
3. Shade the "throw away" side for each linear inequality. The feasible region is the region with no shading!
4. Find all the corner points of the feasible region.
5. Test the objective function at each of the corner points. Choose the best from this list; the largest objective is the maximum, the smallest is the minimum.

Sensitivity of Solutions

- Slack variables tell us how much is left over in the corresponding constraints.

- Shadow price tell us how much the objective changes if we change the right-hand-side of one of the constraints by one.

- Right-hand-side ranges tell us how much we can change the constraints without changing the shadow prices.

- Coefficient ranges tell us how much we can change the numbers in the objective without changing the final values of the variables.

Integer Programs

- Here's the idea behind branch and bound:

  1 Solve the corresponding LP problem (by hand or on the computer). If the variables that need to be integer are, stop.
  2 If a variable that needs to be integer is fractional, "branch" on that variable by creating 2 new sub-problems: If the variable is x and the value is A:
     • create sub-problem 1 by adding the constraint x ≤ (round A down)
     • create sub-problem 2 by adding the constraint x ≥ (round A up)
  3 Solve these new problems; if integer answers come up, choose the best. If a variable which should be integer is not, branch again on that variable.
  4 Repeat this process until the best integer answer is found.

Summary Exercises

In each of the problems below, solve graphically (if possible) and on the computer. Explain the sensitivity information contained in the shadow prices, slack/surplus values, objective coefficient ranges and right hand side ranges.

1)	A factory is set up to make two kinds of greeting cards. A get well card takes 4 minutes to print, 3 minutes to cut and 8 minutes to package. A birthday card takes 5 minutes to print, 2 minutes to cut and 4 minutes to package. The factory's profit is $0.30 and $0.40 resp., for the cards. In the past at least twice as many birthday cards as get well cards have been sold per week, and that is expected to continue. How many of each should be made to maximize profit if there are 200 hours of printing time, 50 hours of cutting time and 100 hours of packaging time available per week?

2) A certain area of forest is populated by two species of animals, eagles and foxes. The forest supplies three kinds of food: fish, squirrels, and mice. For one year, an eagle requires 1 unit of fish, 2 units of squirrel, and 2 units of mice, whereas a fox requires 1 units of fish, 1.2 units of squirrel, and 0.6 units of mice. The forest can normally supply at most 480 units of fish, 720 units of squirrels, and 600 units of mice per year.

a) What is the maximum total number of eagles and foxes that the forest can support?

b) If an eagle is valued at $175 and a fox is valued at $100, how many animals of each species will maximize the value of the animals?

c) If there is a dry spring, then the supply of fish becomes at most 600 units. What is the maximum number of animals that the forest can support now?

3) Francis has identified four attractive stocks in which she wishes to invest her inheritance of $150,000. The first stock is low-risk and is forecasted to return 5% in interest each year. The second stock is also low risk, and is guaranteed to return 5.5% in interest but has a required minimum investment of $40,000. The third stock is high risk, but has a projected return of 11%. The fourth stock is medium risk and has a projected return of 7.25%. Francis has been advised to have no more than 25% of her portfolio in high-risk stocks, and to have at least twice as much in medium-risk as low-risk stocks. How should she invest her money?

4) A candy store sells two different assortments of mixed chocolate covered nut clusters, each containing varying amounts of peanuts, pecans and almonds. The required percentages of each nut in each mixture is given as:

Assortment	Requirements	Price per pound
Regular	$\leq 40\%$ peanuts $\geq 20\%$ pecans	$4.50
Deluxe	$\leq 25\%$ peanuts $\geq 30\%$ pecans $\geq 30\%$ almonds	$6.00

The cost per pound and maximum available for each nut is: 2000 pounds of peanuts at $1.39, 1700 pounds of pecans at $3.00, and 1550 pounds of almonds at $3.49. How many of each nut should be put in each mix for maximum profit?

5) Boone Wood Products manufactures 4 types of door/door frame combinations: deluxe exterior, deluxe interior, standard exterior, standard interior. Each of these doors requires 4 operations as part of the manufacturing process. The company has no warehouse facilities, so finished products are shipped directly to customers. The various departments got together and collected the data below in order to come up with a production schedule (how many of each door to make):

SALES DEPARTMENT

product	max sales/month	$ price/unit
1: standard interior	100	42
2: deluxe interior	50	60
3: deluxe exterior	70	160
4: standard exterior	30	99

PRODUCTION DEPARTMENT

production requirements in hours per door

product	center 1	center 2	center 3	center 4	$ cost /unit
1	3	8	2	6	25
2	4	3	1	0	40
3	2	1	3	4	100
4	10	8	8	6	54
capacity month	1400	1000	626.25	2000	

RAW MATERIALS ACQUISITION DEPARTMENT

Stock requirements in ft^2 of wood per door

product	1/16" wood	1/8" wood	1/4" wood
1	4	2	0
2	7	4	4
3	1	1	6
4	6	0	4
stock availability/month	677.5	1200	600

a) What should the company produce?

b) Production would like to cut back the hours in one of the centers by 25 hours. Which one should they cut back?

c) Raw Materials can get more of one of the sizes of wood paneling (1/16, 1/8, 1/4). Which should they buy more of?

d) Sales wants to find out how restrictive the price per unit for the deluxe exterior is. Could they raise or lower it $10 without changing the production schedule?

e) The company decides to donate unused wood to the Boone area Habitats for Humanity. How many and what kinds of wood will be available to donate?

6) Wilkes-Barre produces 500 tons of waste per day, and Kingston produces 400 tons of waste per day. Waste must be incinerated at the Noxen or Luzerne incinerators, and each incinerator can process up to 500 tons of waste per day. The cost to incinerate waste is $40/ton at the Noxen incinerator and $30/ton at the Luzerne incinerator. Incineration reduces each ton of waste to 0.2 tons of debris, which must be dumped into one of two landfills, in Nanticoke and Berwick. Each landfill can receive at most two hundred tons of debris per day. It costs $3 per mile to transport a ton of material (either waste or debris). Distances (in miles) between locations are shown in the table:

	Noxen Incinerator	Luzerne Incinerator
Wilkes-Barre	30	5
Kingston	36	42

	Nanticoke Landfill	Berwick Landfill
Noxen Incinerator	5	8
Luzerne Incinerator	9	6

Find the appropriate schedule that will minimize total cost of disposing of waste for Wilkes-Barre and Kingston.

7) The R&D division of Boone Manufacturing has developed three new products to manufacture and plans to try at most two of them. The company needs to decide which 2 to make, how many of each to make and which one of the two factories to make them in. Here are the data:

Hours needed to manufacture each product in each plant

		Product		
	1	2	3	Hours/Week
Plant 1	3	4	2	30
Plant 2	4	6	2	40

Profit per unit and maximum sales potential for each proposed product:

		Product	
	1	2	3
Profit	5000	7000	3000
Sales Potential	7	5	9

SELECTED SOLUTIONS

You Try It Solutions

1.1)   C = 65°   A = 66°   B = 47°

1.2) $4^2 + 8^2 = c^2$ $x^2 + 7^2 = 25^2$ $w^2 + 9.2^2 = 13.6^2$
    a) c = 8.94 cm   a) x = 24 m   a) w = 10.02 ft
    b) 9 cm   b) 24 m   b) 10 ft

1.3)   M = 36.4°   m = 3.12 in   X = 67.69°   y = 107.53 cm

1.4) $200^2 + AC^2 = 500^2$ AC = 458.26 ft

1.5)   a) Total area =   BR wing + FAM/KIT + B'FAST
$$33 \times 26 + 29 \times 17 + \frac{1}{2} 12 \times h$$

where h = altitude of the breakfast nook triangle. We use the
Pythagorean theorem to find h:
$$6^2 + h^2 = 12^2$$
$$h = 10.3923$$

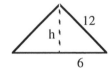

Thus the total area is 858 + 493 + 62.35 = 1413.35, and the cost of
the house is (60)(1413.35) = $84,801.00.

b) Hardwood floors add: (10)(493 + 62.35) = $5,553.50.

2.1) $\tan 67° = \frac{x}{40}$ $x = 40 \tan 67° \approx 94.23$ feet

2.2) $\tan X = \frac{5.6}{9.97} = 0.5617$ $\tan Z = \frac{9.97}{5.6} = 1.7804$

    $\sin X = \frac{5.6}{11.44} = 0.4895$ $\sin Z = \frac{9.97}{11.44} = 0.8715$

    $\cos X = \frac{9.97}{11.44} = 0.8715$ $\cos Z = \frac{5.6}{11.44} = 0.4895$

    $\tan M = \frac{m}{n}$ $\sin M = \frac{m}{p}$ $\cos M = \frac{n}{p}$

Selected Solutions

$$\tan N = \frac{n}{m} \qquad\qquad \sin N = \frac{n}{p} \qquad\qquad \cos N = \frac{m}{p}$$

2.3) $\tan 63.8° = \dfrac{RS}{200}$ $\quad RS = 406.45'$ $\quad \cos 63.8° = \dfrac{200}{ST}$ $\quad ST = 453.00'$

2.4) $\sin 70° = \dfrac{h}{50}$ $\quad h = 46.98'$

2.5) $A = 38.31°$ $\qquad B = 83.00°$ $\qquad\qquad M = 60.46$ $\qquad R = 82.53°$

2.6) $\cos A = \dfrac{10}{32}$ $\quad A = 71.79°$

2.7) $\sin A = \dfrac{4}{18}$ $\quad A = 12.84°$

2.8) $\tan 18.3 = \dfrac{150}{d}$
      $d = 453.56'$

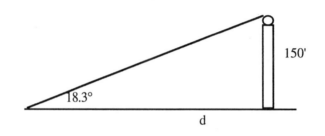

2.9) $A + 18.3° = 90°$
      $A = 71.7°$

      $\tan 71.7 = \dfrac{d}{150}$
      $d = 454'$

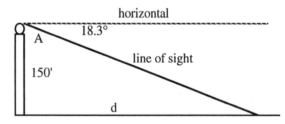

YOU TRY IT SOLUTIONS
TRIGONOMETRY SECTION 1.3

3.1) $\tan 53.4° = \dfrac{h}{100}$
      $h = 134.65$

      $\tan 58.9° = \dfrac{H}{100}$

      $H = 165.77$

      $H - h = 31.1$ ft.

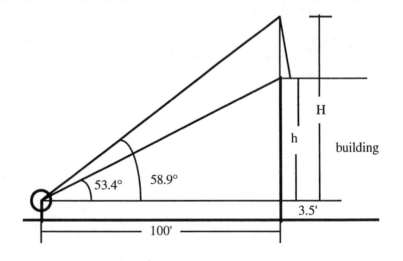

3.2)   Look at the diagram for placement of the angles.

$$\tan 12° = \frac{h}{3+x}$$ $$\tan 18° = \frac{h}{x}$$

$$0.2126 = \frac{h}{3+x}$$ $$0.3249x = h$$

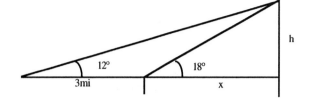

Substitute the value for h on the right into
the equation on the left.

$$0.2126 = \frac{0.3249x}{3+x}$$

Solving for x: $$x = 5.675 \text{ mi}$$
Now back substitute: $$0.3249x = h$$
                            $$1.843945 \text{ mi} = h$$
convert to feet: $$1.843945 (5280) \approx \textbf{9736 feet}$$

3.3)   Look at the diagram for placement of the angles. We'll do part b first.

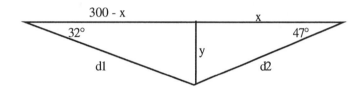

b) $$\tan 32 = \frac{y}{300-x}$$ $$\tan 47 = \frac{y}{x}$$
       $$0.624869 (300 - x) = y$$ $$1.072369 x = y$$

$$0.624869 (300 - x) = 1.072369 x$$
$$x = 110.45$$
$$y = 1.072369 x = 118.4436 \text{ feet (depth of canyon)}$$

a) $$\sin 32 = \frac{y}{d1} = \frac{118.4436}{d1}$$ $$\sin 47 = \frac{y}{d2} = \frac{118.4436}{d2}$$

       $$d1 = 223.51 \text{ feet}$$ $$d2 = 161.95 \text{ feet}$$

3.4)   Look at the diagram for placement of the angles. We are looking for x.

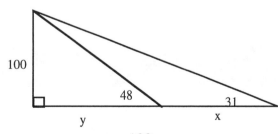

$$\tan 48 = \frac{100}{y}, \text{ so } y = 90.0404.$$

$$\tan 31 = \frac{100}{x+y} = \frac{100}{x+90.0404}, \text{ so } x + 90.0404 = 166.42795$$

$$x = 76.39 \text{ feet.}$$

Selected Solutions

4.1) The pool area is: $\pi R^2 = \pi(12)^2 = 452.38934$ sq ft

The area of the patio and the pool together is: $\pi R^2 = \pi(18)^2 = 1017.87602$ sq ft

The difference is 565.4867 sq ft. This is the area of the patio. Now we need to find the volume by multiplying by the thickness which is 8 inches or 8/12 feet (to get all the units the same).

$$\text{volume} = 565.4867 * 8/12 = 376.99 \text{ cu ft.}$$

Our last step is to calculate the price -- but there's one complication: we have the price per cubic yard, so we'll have to change cubic feet to cubic yards: there are 3 feet in a yard, so there must be 3^3 cubic feet in a cubic yard.

$$\text{price} = 376.99 / 3^3 * \$30 = \$418.88$$

4.2) Let's find the amount of corn first: Area of the base of the silo $= \pi R^2 = \pi(5)^2 = 25\pi$ sq ft.

The volume of the silo (at 2/3 full) must be $25\pi *$ height $= 25\pi * 2/3 (30) = 1570.8$ cu ft.

Now the hay: The volume of the barn is $30 * 25 * 20 = 15000$ cu ft. The volume of one bale of hay is $2 * 1.5 * 4 = 12$ cu ft. How many bales should fit?

volume of barn / volume of a hay bale = 15000 / 12 = 1250 bales of hay.

1.1) microwave: 89(1.05) = \$93.45   tv: 249(1.05) = \$261.45
stereo: 68.88(1.05) = \$72.32   cruise: 2105(1.05) = \$2210.25

1.2) standard: x(1.15) = 103.50, so x = \$90
professional: x(1.15) = 126.50, so x = \$110
competition: x(1.20) = 174, so x = \$145

1.3) 100000(0.02) = \$2,000

1.4) 1000(0.116) + 1000(0.092) + 1000(0.0525) + 1000(0.07) + 1000(0.08) = 410.50

2.1) a) $A = 10,000(1 + 0.09)^5 = \$15,386.24$
   b) $A = 10,000(1 + 0.095)^5 = \$15,742.39$ - \$356.15 more with the 9.5%
   c)   at 9%:   15,386.24 - 10,000 = \$5,386.24
        at 9.5%   15,742.39 - 10,000 = \$5,742.39

2.2) a) n = 40 $r = \dfrac{0.12}{4} = 0.03$ $A = 850(1 + 0.03)^{40} = \$2772.73$

   b) n = 40 $r = \dfrac{0.125}{4} = 0.03125$ $A = 850(1 + 0.03125)^{40} = \$2910.57$

296

difference: $\quad$ 2910.57 - 2772.73 = \$137.84

c) \$1922.73 at 12% $\quad$ and $\qquad$ \$2060.57 at 12.5%

2.3) $\quad$ a) $\quad$ n = 15 $\qquad$ r = 0.068 $\qquad$ $A = \dfrac{850[(1+0.068)^{15} - 1]}{0.068} = \$21{,}033.49$

b) $\quad$ she deposited \$850 once a year for 15 years: $\quad$ 850 x 15 = \$12,750
the balance of the value is interest: $\quad$ 21033.49 - 12750 = \$8,283.49

c) $\quad$ $A = \dfrac{850[(1+0.068)^{1} - 1]}{0.068} = \$850$

2.4) $\quad$ a) $\quad$ n = (12)(2) = 24 $\quad$ $r = \dfrac{0.0875}{12} = 0.0072917$

$$A = \dfrac{110[(1+0.0072917)^{24} - 1]}{0.0072917} = \$2873.68$$

b) $\quad$ he makes 24 deposits of \$110, so the total deposit is: $\quad$ 24x110 = 2640
the rest must come from interest: $\quad$ 2873.68 - 2640 = \$233.68

c) $\quad$ n = (12)(2.5) = 30 $\quad$ $(2\frac{1}{2}$ years = 30 months$)$ $\qquad$ r = 0.0072917

$$A = \dfrac{110[(1+0.0072017)^{30} - 1]}{0.0072917} = \$3673.87$$

2.5) $\quad$ a) $\quad$ n = (12)(2) = 24 $\qquad$ $r = \dfrac{0.06}{12} = 0.005$ $\qquad$ A = 2500 $\qquad$ R = ??

$$2500 = \dfrac{R[(1+0.005)^{24} - 1]}{0.005}$$

$$12.5 = R(0.1271598)$$

$$\$98.30 = R$$

he deposits \$98.30 a total of 24 times: $\quad$ 98.30x24 = \$2359.20

b) $\quad$ n = (12)(4) = 48 $\qquad$ r = 0.005 $\qquad$ $2500 = \dfrac{R[(1+0.005)^{48} - 1]}{0.005}$

$$12.5 = R(0.2704892)$$

$$\$46.21 = R$$

he deposits \$46.21 a total of 48 times: $\quad$ (46.21)(48) = \$2218.08

c) $\quad$ Even though the deposits are smaller, the longer time period gives more opportunity for funds to draw interest. Thus you need less money to achieve the same results.

2.6) $\quad$ a) $\quad$ n = 1095 $\qquad$ $r = \dfrac{0.04}{365} = 0.0001096$ $\qquad$ $4000 = P(1 + 0.0001096)^{1095}$

$\qquad$ P = \$3547.66

b) $\quad$ n = 1095 $\qquad$ $r = \dfrac{0.05}{365} = 0.000137$ $\qquad$ $4000 = P(1 + 0.000137)^{1095}$

$\qquad$ P = \$3442.82

CONCLUSION: The 5% account requires a smaller deposit, but still yields the \$4000 in three years. Even after taking into consideration the extra \$10 fee for a new account at the 5% bank, John would still need \$94.84 less **now** to be able to buy his \$4000 car in three years. So send John to the 5% bank.

Selected Solutions

2.7)   a) $A = \dfrac{25[(1+0.0125)^{40} - 1]}{0.0125} = 1287.24$ b) $A = \dfrac{500[(1+0.05)^{25} - 1]}{0.05} = 23863.55$

$1287.24 = P(1 + 0.0125)^{40}$ $23863.55 = P(1 + 0.05)^{25}$

$P = \$783.17$ $P = \$7046.97$

2.8) a) Applying logs to both sides: $\log(2.15) = n \log(1.000124)$

$$\frac{\log(2.15)}{\log(1.000124)} = n, \text{ so } n = 6173.51.$$

b) Let's simplify what's in the base first: $\left(1 + \dfrac{.075}{4}\right) = 1.01875$

Applying logs to both sides: $\log(4.1) = n \log(1.01875)$

$$\frac{\log(4.1)}{\log(1.01875)} = n, \text{ so } n = 75.96.$$

c) We need to isolate the exponent part before applying the log rule:

$$1000 = \frac{125(1.11^{n} - 1)}{.11}$$

$$\frac{1000 \times .11}{125} = 1.11^{n} - 1$$

$$.88 = 1.11^{n} - 1$$

$$1.88 = 1.11^{n}$$

Applying logs to both sides: $\log(1.88) = n \log(1.11)$

$$\frac{\log(1.88)}{\log(1.11)} = n, \text{ so } n = 6.05.$$

2.9) $r = \dfrac{0.08}{4} = 0.02$ $12,000 = 8500(1 + 0.02)^{n}$

$\dfrac{12000}{8500} = 1.02^{n}$

$1.4117647 = 1.02^{n}$

$\ln 1.4227647 = \ln 1.02^{n}$

$\ln 1.4227647 = n(\ln 1.02)$

$0.3448405 = n(0.0198026)$

$17.41 = n$ approximately

2.10)   a) $n = ??$ $r = \dfrac{0.085}{4} = 0.02125$ $A = 12,000$ $R = 650$

$12,000 = \dfrac{650[(1+0.02125)^{n} - 1]}{0.02125}$

$255 = 650[(1+0.02125)^{n} - 1]$

$0.3923077 = (1.02125)^{n} - 1$

$1.3923077 = 1.02125^{n}$

298

$$\ln\ 1.3923077 = n(\ln 1.02125)$$

$$\frac{\ln\ 1.3923077}{\ln 1.02125} = n$$

$$15.74 = n \quad \text{or} \quad \text{about 16 quarterly deposits or 4 years}$$

b) We could do the problem using the new r and R values, then solve for n. But since the question we are trying to answer is whether part b would get the down payment "sooner or later" than part a, we could rephrase the question this way:

Suppose deposits of $600 are made into a similar account paying 9% for that same 4 years--would we have MORE than the $12,000 needed **OR** would we have LESS than the $12,000?

using: $n = (4)(4) = 16$ $A = \dfrac{600[(1+0.0225)^{16} - 1]}{0.0225} = \$11,403.24$

YOU TRY IT SOLUTIONS
FINANCE SECTION 2.3

3.1) a) $n = (4)(2) = 8$ $r = \dfrac{0.12}{4} = 0.03$ $R = \dfrac{(4300)(0.03)}{[1 - (1+0.03)^{-8}]} = \$612.56$

b) 8 payments of 612.56 yields: $(8)(612.56) = \$4900.48$

3.2) a) after the down payment of 1500, Dave owes $13899-1500 = \$12,399$

$n = 12 \times 3 = 36$ $r = \dfrac{0.126}{12} = 0.0105$ $R = \dfrac{(12399)(0.0105)}{[1 - (1+0.0105)^{-36}]} = \$415.39$

b) 36 payments of 415.39 results in payments of $(36)(415.39) = \$14,954.04$

plus the down payment of $1500: $\$14,954.04 + \$1500 = \$16,454.04$

Interest = total paid - cost of car: $I = 16454.04 - 13899 = \$2,555.04$

c) $n = (12)(1.5) = 18$ $r = 0.0105$ $R = \dfrac{(12399)(0.0105)}{[1 - (1+0.0105)^{-18}]} = \$759.58$

The payment for a 3 year loan is $415.39 while the payment for 1.5 years is $759.58.
Interest for 3 year loan: $2555.04   Interest for 1.8 year loan: $1273.44
Total paid using 3 yr loan: $16454.04   Total paid using 1.8 yr loan: $15172.44

3.3)

Payment Number	Payment Amount	Portion to Interest	Portion to Principal	Balance Due
0	0	0	0	$2,000
1	804.23	.10x2,000 = 200	804.23-200 = 604.23	2,000-604.23 = 1395.77
2	804.23	.10x1395.77 = 139.58	804.23-139.58 = 664.65	1395.77-664.65 = 731.12
3	804.23	.10x731.12 = 73.11	804.23-73.11 = 731.12	731.12-731.12 = 0

3.4)

Payment Number	Payment Amount	Portion to Interest	Portion to Principal	Balance Due
0	0	0	0	$14,500
1	381.84	.01x14,500 = 145.00	381.84-145 = 236.84	14500-236.84 = 14263.16
2	381.84	.01x14263.16 = 142.63	381.84-142.63 = 239.21	14263.16-239.21 = 14023.95
3	381.84	.01x14023.95 = 140.24	381.84-140.24 = 241.60	14023.95-241.60 = 13782.35
4	381.84	.01x13782.35 = 137.82	381.84-137.82 = 244.02	13782.35-244.02 = 13538.33

3.5)

Payment Number	Payment Amount	Portion to Interest	Portion to Principal	Balance Due
0	0	0	0	297,000
1	85,711.67	17,820.00	67,891.67	229,108.33
2	85,711.67	13,746.50	71,965.17	157,143.16
3	85,711.67	9,248.59	76,283.08	80,860.08
4	85,711.68	4,851.60	80,860.08	0

YOU TRY IT SOLUTIONS
FINANCE SECTION 2.4

4.1)

DEBT		INCOME	
Rent:	375.00	Teaching:	20,400.00
Fab Furn:	22.50	Summer:	3,000.00
Insurance	50.00	TOTAL:	23,400.00 ÷ 12 mo
TOTAL	447.50		= 1950.00

Debt to income ratio: $\dfrac{447.50}{1950} = 0.2295 \approx 23\%$ current ratio

Maximum Debt (40%): $\dfrac{DEBT}{1950} = 0.40$ DEBT = $780

Linda's maximum payment (40%): 780 - 447.50 = **$332.50**

4.2) 48 months: $247 = \dfrac{P(.0075)}{1 - (1+.0075)^{-48}}$ solving, P = $9925.64

Adding the $2800 down payment, the new car should sell for less than **$12,725.64**

60 months: $247 = \dfrac{P(.0075)}{1 - (1+.0075)^{-60}}$ solving, P = $11,898.82

Adding the $2800 down payment, the new car should sell for less than **$14,698.82**

4.3) $247 - 12.50 = \$234.50$ $234.50 = \dfrac{P(.0075)}{1 - (1+.0075)^{-36}}$ solving, $P = \$7374.28$

Adding the $2800 down payment, the new car should sell for less than **$10,174.28**

YOU TRY IT SOLUTIONS
FINANCE SECTION 2.5

5.1)

High	Low	Stock	Div	P/E	100s	High	Low	Last	Change
$20\frac{1}{4}$	$13\frac{1}{4}$	ARG	4.10	30.8	35	$15\frac{5}{16}$	15	$15\frac{3}{8}$	$-\frac{1}{8}$

   a) closed today at 15.38
   b) overvalued since 30.5 >>> 7
   c) 3500 shares were traded
   d) 4.10 / 15.38 = 26.66% yield
   e) varied between 13.25 and 20.25 in the last year; varied 15 and 15.31 in the last day.

YOU TRY IT SOLUTIONS
STATISTICS SECTION 3.1

1.1) a) ASU incoming freshmen, SAT   b) Students at ASU, GPA
     c) North Carolinians, Pepsi vs Coke   d) Goodyear tires, Lifetime

1.2) a) only gets students who eat in cafeteria - too convenient
     b) lobby in only one building - too convenient
     c) voluntary response - only people who feel strongly will bother
     d) 10 block area - does she ask any farmers?

1.3) Numbering the cars 00 to 49 and starting in row 1 of the table gives:
        38   45   49   29   00   12   27   19

1.4) Question 1 is okay; question 3 needs significant rewriting.

YOU TRY IT SOLUTIONS
STATISTICS SECTION 3.2

2.1)

CLASSES	FREQUENCY
$10 \leq$ final exam scores < 20	1
$20 \leq$ final exam scores < 30	0
$30 \leq$ final exam scores < 40	0
$40 \leq$ final exam scores < 50	0
$50 \leq$ final exam scores < 60	10
$60 \leq$ final exam scores < 70	0
$70 \leq$ final exam scores < 80	11
$80 \leq$ final exam scores < 90	2
$90 \leq$ final exam scores < 100	1

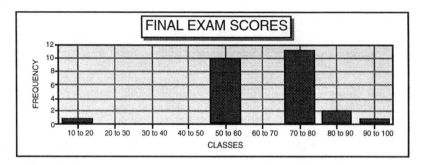

2.2) First: Histogram, second: Relative Frequency Histogram

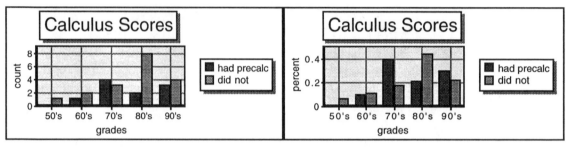

The relative frequency histogram is better; it allows for the number of students to be different between the categories.

2.3) To draw the population data line graph, we put years on the horizontal and population on the vertical:

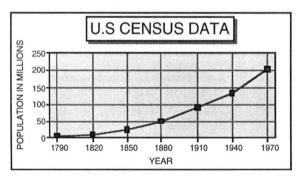

2.4) a) 25 students took the quiz.

b)

CLASSES	FREQUENCY	RELATIVE FREQUENCY
30 to 40	1	1/25
40 to 50	0	0/25
50 to 60	3	3/25
60 to 70	3	3/25
70 to 80	5	5/25
80 to 90	6	6/25
90 to 100	7	7/25

c) There are 18 grades ≥ 70. d)

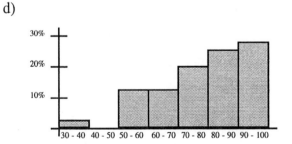

2.5) a) average = (80 + 70 + 60 + 80) / 4 = 72.5, or $72,500.

   b) 1982 is 30 higher than 1981, and the rest are smaller differences, so 1982.

   c) sales / earnings = 3 / 60 = 1 / 20 = 0.05 = 5%.

   d) It makes sense that the costs should be sales - earnings (money in - money kept), so costs = 80 - 8 = 72, or $72,000.

2.6)

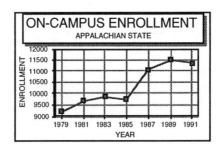

2.7)

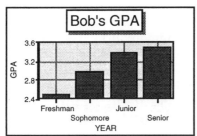 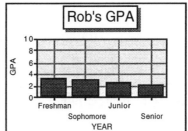

YOU TRY IT SOLUTIONS
STATISTICS SECTION 3.3

3.1)
```
1| 0    2    4    5    5    8   The distribution is skewed right.
2| 0    2    4    5    8
3| 5    8
4| 0    3    5
5| 5
6| 5
7| 5
8| 0
```

3.2)   COMPANY A: median = $24,000   mean = (4*24,000 + 250,000)/5 = $69,200
       COMPANY B: median = $36,000   mean = (5*36,000)/5 = $36,000

   For Company B, both measures are the same. For Company A, the median would best typify an employee's salary.

3.3)   For both males and females, n = 10.
```
Males Ordered:    10   14   15   15   18   28   38   40   55 80
Females Ordered: 12   20   22   24   25   35   43   45   65 75
                           Q1         M         Q3
```

   Five Number Summary (Males)   Five Number Summary (Females)
   LO = 10   LO = 12
   Q1 = 15   Q1 = 22
   M = 23   M = 30
   Q3 = 40   Q3 = 45
   HI = 80   HI = 75

3.4) a) between $35,000 and $44,000

b) between $25,000 and $39,000

c) Engineering. 75% make $35,000 or above. The upper 75% of Agriculture salaries are only $25,000 or above. The Engineering Median is also higher .

d) The upper 25% of Engineering students make between about $44,000 and $50,000. The upper 25% of Agriculture students make between about $39,000 and $53,000. There is more variability in the upper end of the Agriculture data. The highest salaries, however, would go to the Agriculture students.

e) about $22,000 to $25,000   f) Agriculture shows more variability.

YOU TRY IT SOLUTIONS
STATISTICS SECTION 3.4

4.1) a) $y - 310 = (80/11)(x - 48)$
$y = (80/11)x - 3840/11 + 310$
$y = 7.27x - 39.09$

b) First find the slope m: (rounding to nearest hundredth)

$$m = \frac{35 - 52}{140 - 210} = \frac{-17}{-70} = 0.24$$

Write the equation for the line using either point; here, we will use (140,35), rounding to nearest hundredth: $y - 35 = 0.24(x - 140)$
$$y = 0.24x + 1.4$$

4.2) Here's a scatter plot of year versus CPI with a line of best fit drawn in.

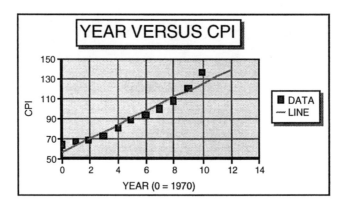

1) We want to use the graph to check on 1981 - that's year 11. Read up from 11 to the line and across gives a CPI = 132.

2) We want to find an equation for our line. It looks like it goes through (2,70) and 5,90), so the line will be

$$y - 70 = \frac{20}{3}(x - 2).$$

Plugging 11 for x (that's 1981) gives y = 130.

4.3) We substitute x = 95 into the regression line $y = 6.768x - 18.295$:
$$y = 624.665,$$

or $624,665 in sales. This could be an unreliable prediction for several reasons: The test score of 95 is extreme in comparison to the other test scores in the data. This could mean that a test score of 95 is not even possible - maybe a score of 50 is 100%! Furthermore, the sales projection of $624,665 appears very unreasonable. It is approximately a 70% increase over the highest projection in the data set ($360,000). While this may not be impossible, it is quite unlikely. There is an upper limit to reasonable sales projection figures for any salesman, due to market constraints, resource allocations, production capabilities, etc.

4.4) The picture is already drawn above in 4.2.

a) regression line is $y = 6.85 x + 56.7$. $r = 0.9754$; $r^2 = 95\%$ - that's a strong correlation.

b) predicted CPI $= 6.85 (11) + 56.7 = 132.05$

YOU TRY IT SOLUTIONS
STATISTICS SECTION 3.5

5.1) Clearly the numbers are not linear, so let's look at the new/old ratios:

$$8332/7179 = 1.1606$$
$$8749/8332 = 1.0500$$
$$11498/8749 = 1.3142$$
$$13741/11498 = 1.1950$$
$$16632/13741 = 1.2104$$
$$20353/16632 = 1.2237$$
$$25130/20353 = 1.2347$$
$$30844/25130 = 1.2274$$

At first these are changing, but notice how they settle down to about 1.23. So our 5 year rate is about 23%. What is the one year rate? We need to find r so that

$$(1 + r)^5 = .1 + .23$$

Trial and error shows that $r = 0.043$, or about 4.3% per year.

YOU TRY IT SOLUTIONS
LINEAR PROGRAMMING SECTION 4.1

1.1) A few points:

(B,K)	INEQ1	INEQ2	INEQ3	INEQ4	PROFIT
(2000,1000)	T	T	T	T	$18,500
(10000,1000)	F	F	F	T	not feasible
(8000,2000)	F	F	F	T	not feasible
(9000,4000)	F	F	F	T	not feasible
(5000,2000)	T	T	T	T	$43,000

1.2) 2) S = amount of salad to serve   M = amount of meat to serve

3) minimize calories: minimize $Z = 30 S + 70 M$

4)
protein:	$0.2 S + 0.5 M \geq 3$
fat:	$0.2 S + 0.3 M \geq 1$
carbs:	$0.6 S + 0.2 M \geq 3$

1.3) 2) M = minutes of morning time to buy for 2 weeks
A = minutes of afternoon time to buy for 2 weeks
E = minutes of evening time to buy for 2 weeks

3) maximize exposure: maximize $Z = 200000 M + 100000 A + 600000 E$

4)
budget:	$3000 M + 1000 A + 12000 E \leq 102000$
prime/evening limit:	$E \leq 6$
total limit:	$M + A + E \leq 25$

2.1)

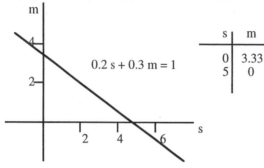

s	m
0	3.33
5	0

$0.2 s + 0.3 m = 1$

x	y
10	0
0	6

2.2)

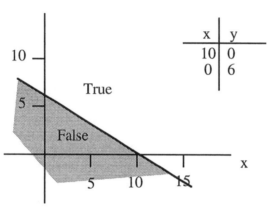

True

False

2.3) A few feasible points:

(B,K):	(5000,2000)	(4000,0)	(3000,1000)
Profit:	$43,000	$24,000	$24,500

2.4) $(x,y) = (\frac{15}{2}, \frac{7}{2})$ $(d,p) = (\frac{144}{89}, -\frac{218}{89})$

2.5)   variables: E = number of Econovacs displayed;   S = number of Supervacs displayed
       objective: maximize profit: maximize $Z = 30E + 50S$
       constraints: sqft: $3E + 4S \leq 61$
                      investment: $80E + 150S \leq 2060$

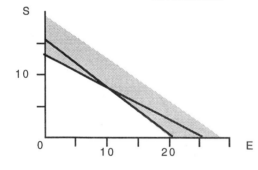

corner point (E,S)	profit
(0,0)	$0
(20.33,0)	$609.90
(0, 13.73)	$686.50
(7,10)	$710 MAXIMUM!

Sally should stock 7 Econovacs and 10 Supervacs to maximize her profit.

2.6)   variables: S = amount of salad to serve   M = amount of meat to serve
       objective: minimize calories: minimize $Z = 30 S + 70 M$
       constraints: protein: $0.2 S + 0.5 M \geq 3$
                           fat: $0.2 S + 0.3 M \geq 1$
                           carbs: $0.6 S + 0.2 M \geq 3$

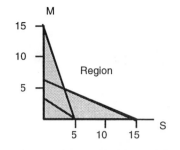

corner point (M,S)	cals	
(0,15)	450	
(15,0)	1050	
(3.46,4.62)	426.9	MINIMUM

YOU TRY IT SOLUTIONS
LINEAR PROGRAMMING SECTION 4.3

3.2) a) The number of Kens and Barbies to make stays the same but the profit rises to $49,900.

b) The number of Kens increases to 2916.67 and the number of Barbies increases to 5833.33. This gives a profit of $53,958.33.

3.4)

	PROFIT	
Inequality	Increase by one	Decrease by one
diamonds	8900	8700
labor	8800	8800
emeralds	8800	8800

We get exactly what we expect: diamond constraint up gives profit up by 100, down gives profit down by 100; labor and emeralds (with shadow prices 0) do not effect the profit.

3.5) If we go past the limits, the entire solution changes. If we are within the limits, the number of necklaces to make stays the same, but the profit changes.

3.6)

Necklace 1 Profit	Solution
$165	N1 = 0, N2 = 33.33, Profit = $8333.33
$1000	N1 = 14, N2 = 24, Profit = $20,000

Necklace 2 Profit	Solution
$299	N1 = 14, N2 = 24, Profit = $9,976.00
$1000	N1 = 0, N2 = 33.33, Profit = $33,333.33

Within the range the same number of necklaces are made, but outside everything changes.

3.7) a) 1,000,000 pounds of wool and 3,333,333 pounds of acrylic fabrics
b) $2,000.00 because S(2) = 100.
c) The amount of each fabric made will stay the same because $35 is within the 32 to infinity range.
d) Add more room for wool. The shadow price (30.0) is the largest shadow price.

3.8) minimize   2.89 b1 + 2.95 b2 + 4.62 b3   (cost)
st.   12 b1 + 10 b2 + 20 b3 ≥ 16   (nitrogen)
            8 b1 + 10 b2 + 2 b3 ≥ 3   (potassium)
            6 b1 + 10 b2 + 2 b3 ≥ 6   (phosphorus)
            8 b1 + 10 b2 + 2 b3 ≤ 6   (maximum potassium)
a) solution: Mix 0.49 parts of b2 and 0.56 parts of b3 for a minimum cost. (This amount gives $4.01 for 1.05 pounds of mix.)
b) slacks: The mix is 16-6-6, there is a surplus of potassium.
c) shadow prices: Nitrogen is worth 22 cents -- if we needed 1 part more, it would cost 22 cents more. Phosphorus is worth 18 cents -- if we needed one part more, it would cost us 18 cents more. Potassium (the at least) we have more than enough of. Potassium (maximum allowable) is worth 11 cents -- if we raise the maximum allowable potassium,

we can lower the cost by 11 cents.

YOU TRY IT SOLUTIONS
LINEAR PROGRAMMING SECTION 4.4

4.1) The fractional solution:
We draw the region as usual, finding the corner points:

x	y	Z (8x+5y)
0	0	0
5	0	40
0	6	30
3.75	2.25	41.25

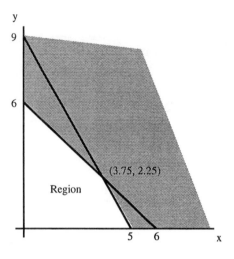

Now let's find all of the integer feasible points by looking at each x and finding all the y's that would satisfy the constraints:

x = 0: y = 0, y = 1, y = 2, y = 3, y = 4, y = 5, y = 6
x = 1: y = 0, y = 1, y = 2, y = 3, y = 4, y = 5
x = 2: y = 0, y = 1, y = 2, y = 3, y = 4
x = 3: y = 0, y = 1, y = 2, y = 3
x = 4: y = 0, y = 1
x = 5: y = 0

Checking all of these pairs shows that the best integer solution is: x = 5, y = 0, Z = 40.

4.2) Solving this problem gives the regions for the branches::

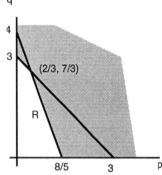

Optimum occurs at p = 2/3 and q = 7/3; Z = 11/3
so we need to branch on p

p ≤ 0   p ≥ 1

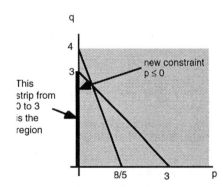

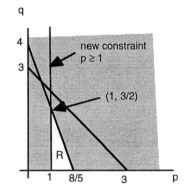

the optimal answer here
is p = 0, q = 3, and Z = 3

the optimal answer here
is p = 1, q = 3/2 and Z = 7/2

So the right branch wins and $p = 1$, $q = 3/2$ and $Z = 7/2$.

4.3)   Setting up the problem, following the hint:

minimize $40 F1 + 40 F2 + 15 P1 + 15 P2 + 15 P3 + 15 P4 + 15 P5 + 15 P6$
subject to:
$$P1 + P2 + P3 + P4 + P5 + P6 \le 5$$
$$F1 + F2 + P1 \ge 4$$
$$F1 + F2 + P1 + P2 \ge 3$$
$$F1 + F2 + P1 + P2 + P3 \ge 4$$
$$F2 + P2 + P3 + P4 \ge 6$$
$$F1 + P3 + P4 + P5 \ge 5$$
$$F1 + F2 + P4 + P5 + P6 \ge 6$$
$$F1 + F2 + P5 + P6 \ge 8$$
$$F1 + F2 + P6 \ge 8$$
all variables integers ≥ 0

Here is the tree for the integer program solution:

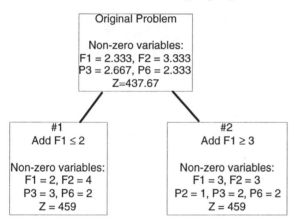

We have two equal integer answers to choose from!

Solution 1: Hire 2 full time noon-lunchers, 4 full time 1-lunchers, 3 part timers starting at   11 and 2 part timers starting at 2 for a cost of $459.

Solution 2: Hire 3 full time noon-lunchers, 3 full time 1-lunchers, 1 part timer starting at   10, 2 part timers starting at 11 and 2 part timers starting at 2 for a cost of $459.

4.4)   Let the investments be I1, I2, I3, and I4:

maximize $16 I1 + 22 I2 + 10 I3 + 9 I4$
subject to:
$$5 I1 + 7.5 I2 + 4 I3 + 2.5 I4 \le 14$$
and the if then:
$$I3 \ge I2$$
all variables binary

Here's the tree:

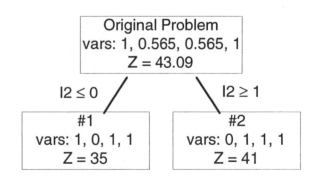

We have two integer solutions to choose from; we'll choose the best - in this case the biggest - so Janet should invest in #2, #3 and #4 for a net present value of $41,000.

END OF SECTION EXERCISES

END OF SECTION EXERCISES
TRIGONOMETRY SECTION 1.1

1) 55°

3) X = 90°   Z = 17.9°   x = 19.95 m

5) 14.32 feet

7) 31.65 m

9) 196.88 ft

11) $1860.06

13) 97.20 cu ft

15) 179.66°

17) C

19) D

21) C

23) C

25) C

27) A

29) A

31) C

END OF SECTION EXERCISES
TRIGONOMETRY SECTION 1.2

1) A = 62.96°   B = 90°   C = 27.04°   a = 754.44 ft

3) A = 14.84°   B = 90°   C = 75.16°   b = 398.28 ft

5) 79.36 ft

7) 30.07 ft

9) 86.05 m

11) 9659.29 ft

13) 78.18 m

15) 350.41 ft

17) 38.68°

19) a) 7.49 ft
     b) 1.52 ft

21) 12,375.43 ft

23) 7.69°

25) a) 375 ft
     b) 51.34°

END OF SECTION EXERCISES
TRIGONOMETRY SECTION 1.3

1) 21.07 ft

3) 9949 ft

5) 12951 ft

7) 104.62 ft

9) 20.03 m

11) 110.49 ft

13) 51.59 ft (pole)   distances: 110.64' and 89.36'

15) 194.85 ft

17) slant length is just over 143 ft
     144 gives 12240 sq ft of carpet

19) 22.77 ft

END OF SECTION EXERCISES
TRIGONOMETRY SECTION 1.4

1a) p = 6.71
     a = 2.36

1b) p = 4.27
     a = 1.13

1c) p = 2.69
     a = 0.21

3) 50.13 ft 5) 3053.6 cu ft 7) 15.71 cu ft
                   $7634.07

9) 1005.31 sq ft to paint; 5.03 or 6 cans

11) A 13) B 15) d

17) b

END OF SECTION EXERCISES
FINANCE SECTION 2.1

1) a) 47 b) 82.5% c) 218.75

3) a) 82.5 b) 53.57% c) 291.67

5) sweater= $134.10 wallet = $44.99 suit = $719.10

7) Yes, flat rate is better.

9) a) 42.5% 11) yes

13) A 15) C 17) C 19) B

END OF SECTION EXERCISES
FINANCE SECTION 2.2

1) $3,951.83 3) $1579.39 5) $5232.45
    $1,651.83   deposit: $900 $1832.45
                   interest: $679.39

7) $7532.38 9) A = $15,924.16 -- YES!

11) n = 24.54 compoundings, which is 25/2 = 12.5 = 12 years 6 months

13) n = 1116.64 ≈ 1117 days or about 1117/365 = 3.06 years

15) $67,121.04 17) 15 years 19) $4,503.86

21) Alice has more; Ed deposits more

23) AFTER TEN YEARS:
     Option 1: (lump sum deposit) A = $170,981.89
     Option 2: (savings plan) A = $157,737.54 Choose Option 1!

END OF SECTION EXERCISES
FINANCE SECTION 2.3

1) $2025.98 3) R = $549.18 5) 5 year: R = $740.99
                   total = $28,660.64   total = $49,259.40
                                           10 year: R = $457.41
                                           total = $59,689.20

7) Amount: $3,100.00 Periodic Rate: 0.14/1 = 0.14
    Number of Payments: 3

311

Payment Number	Payment Amt	Interest	Principal	Balance
0	$0	$0	$0	$3,100.00
1	$1,335.27	$434.00	$901.27	$2,198.73
2	$1,335.27	$307.82	$1,027.45	$1,171.29
3	$1,335.27	$163.98	$1,171.29	$0

9)   Amount: $19,102.50   Periodic Rate: $0.12/12 = 0.01$ Number of Payments: 60

Payment Number	Payment Amt	Interest	Principal	Balance
0	$0	$0	$0	$19,102.50
1	$424.92	$191.03	$233.89	$18,868.61
2	$424.92	$188.69	$236.23	$18,632.38
3	$424.92	$186.32	$238.60	$18,393.78
4	$424.92	$183.94	$240.98	$18,152.80

11)   Amount: $75,000.00   Periodic Rate: $0.084/12 = 0.007$ Number of Payments: 360

Payment Number	Payment Amt	Interest	Principal	Balance
0	$0	$0	$0	$75,000.00
1	$571.38	$525.00	$46.38	$74,953.62
2	$571.38	$524.68	$46.70	$74,906.92
3	$571.38	$524.35	$47.03	$74,859.89

13) a) $6453.38; $8271.67; $9922.13   b) $7953.38; $9771.67; $11422.13

END OF SECTION EXERCISES
FINANCE SECTION 2.4

1)   21.68%

3)   a) 19.36%
     b) $311.91

5)   yes

7) $2966.31

9) $14,344.46

11) a) 27.93%
     b) $199
     c) $8,756.82

13) a) $552.25   b) $558.25   c) $560.53

END OF SECTION EXERCISES
FINANCE SECTION 2.5

1) $19,277.10

3) $3392.82

5)   6.346%

7)   a) $15; $16; -6.25%   b) undervalued   c)   1200
     d) 6.4%   e) yes; no

END OF SECTION EXERCISES
STATISTICS SECTION 3.1

1)   If they tested them all, there would be none left to sell.

3)   a) May not be same type of customers that Madame Whiffit has --too convenient.

b) Possibly only people who *really* support a candidate will respond--voluntary response. Readership may not reflect opinions as a whole--too convenient.

5) a) Population: Black residents of Chicago
        Variables measured: Feelings about police service
        Sample: People willing to do interview.
        Bias: People may be intimidated by police officer conducting the interview.

    b) Population: Minneapolis households
        Variable measured: Usage of their bread products
        Sample: People at home who will do the interview
        Bias: People who work during regular hours will not be at home to participate.

7) Taking rows 1 through 5 gives between 10 and 20 occurrences of each of the digits 0 through 9; alone this is not enough to guarantee randomness. If there is no obvious pattern and the digits occur about the same number of times, we have a pretty good indication though.

9) Answers will vary, but using the random number table and larger samples should give answers closest to the true average of .5

11) Answers will vary.

13) These questions are biased towards supporting legislation,

15) Answers will vary.

17) What about other languages?

END OF SECTION EXERCISES
STATISTICS SECTION 3.2

1) Answers will vary.   3) Answers will vary.   5) Answers will vary.

7) Answers will vary.   9) Answers will vary.

11) a) Approximately 35% without and approximately 60% with reinforcement.

    b) The most dramatic decrease in retention occurs in the first 3-4 hours. (Line is steepest.)

    c) Approximately 40%.

13) a) 24 mpg
    b) 10.4%; 65%

15) a) unskilled   7) B   9) B
    b) skilled and farmers
    c) cannot be determined

END OF SECTION EXERCISES
STATISTICS SECTION 3.3

1) The year 1981 with 22 home runs is an outlier. An approximate center would be 39 home runs.

```
 2  |  2
31  |  2   2   2   3
3h  |  6   6   7   9   9   9
41  |  0   0   1   2   3   4
4h  |  5   6   9
```

3)
```
4h  |  6   9
51  |  3
5h  |  6   6   7   8
61  |  0   0   3   3   4   4
6h  |  5   6   7   7   7   8
71  |  0   1   1   2   3   4
7h  |  7   8   8   9
81  |  0   1   3
8h  |  5   8
91  |  0   0
```

5)   Mean = 68.8, median = 67, mode = 67. The mean is the best approximation.

7)   a) Yes.
      b) Yes.

9)   The mode always appears in the data list.

11)   The five number summary: 93, 95.5, 100, 101.5, 138; the mean would be greater.

13)   The data are roughly symmetric.

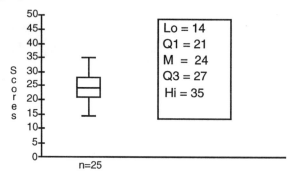

15)   a) The graph is skewed right.   b) 2
      c) 13

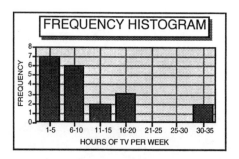

17)   a) Approximately 57%   b) Approximately 46%

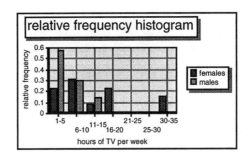

19) Lo=0, Q1=3.5, M=7.5, Q3=19, Hi=37;

a) 7.5; above (mean=11.85)   b) 15.5

c) 30%; 30%; 80%

21) Lo = 51, Q1 = 66, M = 71, Q3 = 80, Hi = 98

a) Approximately 25%, 25%, and 75%.   b) slightly skewed right.

23) SMALL: Lo=24, Q1=26, M=27.5, Q3=29.5, Hi=35
SPORTY: Lo=20, Q1=26, M=28, Q3=29, Hi=32
COMPACT: Lo=21, Q1=23, M=24, Q3=24, Hi=24
MIDSIZE: Lo=17, Q1=19.5, M=21, Q3=21, Hi=23

a) Compact   b) Sporty   c) 92%, 91%, 0%, 0%
d) 3.5; 3; The small cars have a higher variability than the compact cars.

25) a)

	ABC	CBS	NBC
LO	20	22	21
Q1	24	25	23
M	27	25	23
Q3	27	26	24.5
HI	31	36	27

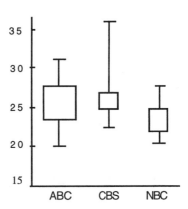

ABC shows did better because 50% of them have higher shares than nearly all CBS shows and all NBC shows. NBC still performed the worst since 75% of CBS shows have higher shares than 75% of NBC shows.

b)

	ABC	CBS	NBC
Lo	13.5	13.1	13.1
Q1	14.4	15.5	13.6
M	16.6	16.35	14.5
Q3	17.2	17.3	15.0
Hi	20.2	21.9	17.6

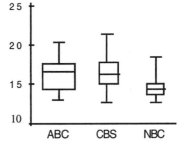

CBS shows have a better rating than ABC. (75% of CBS shows are rated 15.5 or higher while 75% of ABC shows are rated 14.4 or higher.) NBC has the worst performance; 75% of their shows are rated lower than the medians of the others.

END OF SECTION EXERCISES
STATISTICS SECTION 3.4

1)   a) $y = -4x + 86$ b) $y = 2.98x - 1.07$

3)   a) $y = 0.333x + 5.33$ b) $y = -20.6x + 38,799.4$

5)   a) $y = -0.431x + 107.41$ b) $y = 2.6x + 77.5$

7)   a) Scatterplot A: $1.674 + 0.011 + 0.652 + 1.685 + 0.978 + 0.641 = 5.641$
       Scatterplot B: $2 + 0.5 + 0 + 2.5 + 0 + 0.5 = 5.5$
       Therefore, the equation in Scatterplot B is the better fit.

       b) Scatterplot B; the sum of distances from points to the line is smaller.

9)   a) independent variable = body temperature; dependent variable = pulse rate

       b) independent variable = foot length; dependent variable = height

11)   Equation of line connecting $(12,8)$ and $(23,13)$: $y = 0.45x + 2.55$
      Errors predicted: 9 (round to nearest whole error in this context)

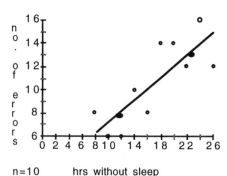

n=10   hrs without sleep

13)   a ---> 3   b ---> 1   c ---> 4   d ---> 2

15)   a) Equation for fitted line depends on the   points chosen, but a reasonable number of
      tickets will be around 35.

      b) approximately -18 tickets per week;

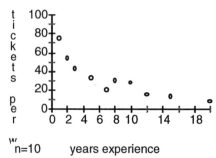

n=10   years experience

      c) The fact that r is negative suggests that the correlation is a decreasing one. The fact that
r^2 is 74% suggests that the relationship is a fairly strong one.

17)   Equation of the line depends on the points chosen, but a reasonable answer will be about 20
      points.

END OF SECTION EXERCISES
STATISTICS SECTION 3.5

1) Very close to exponential; Rate: 3.7% for one year; predicts 10.96

3) Exponential at first, but levels off; Rate: 4.1% for one year early; predicts 44.36

5) Ratio approximately 1.24, so about 24% per year; P = 1000; about 5704 after 8 years; $10,000 after around 11 years

END OF SECTION EXERCISES
LINEAR PROGRAMMING SECTION 4.1

1) a) $P \leq 800$ and $A \leq 600$ b) $A - 3P \geq 0$ c) $4P + 2A \leq 200$

3) a) $16\,ww + 8\,dc + 12\,ds \leq 500$ b) $ww + dc + ds \leq 25$ c) $50\,ds \leq 200$; $150\,dc \leq 1000$

5) C = the number of chocolate bars to serve, G = the number of granola bars to serve
```
Maximize        7C +   2G
subject to      5C + 15G ≥ 40     (protein)
               50C + 25G ≤ 150    (calories)
                C ≥ 0, G ≥ 0
```

7) P = the number of cups of punch to sell, L = the number of cups of lemonade to sell
```
Maximize       10P + 9L
subject to      1P + 1L ≤ 40     (cups)
                5P + 3L ≤ 150    (funds, in cents)
                P ≥ 0, L ≥ 0
```

9) S = the number of shawls to make per week, A = the number of afghans to make per week
```
Maximize       25S + 40A
subject to      1S + 2A ≤ 13     (spinning)
                1S + 1A ≤ 10     (dying)
                1S + 4A ≤ 30     (weaving)
                S ≥ 0, A ≥ 0
```

11) R = number of regular lockers to buy, L = number of large lockers to buy
```
Maximize       10R + 20L
subject to:    40R + 100L ≤ 600   (budget)
                4R +   8L ≤ 52    (space for cabinets)
                R≥ 0, L ≥ 0
```

13) R = pounds of regular, D = pounds of deluxe, S = pounds of supreme
```
Maximize        4R +    4D +  4S
subject to     .50R + .30D + .20S ≤ 800    (pounds of peanuts)
               .30R + .40D + .40S ≤ 400    (pounds of cashews)
               .20R + .30D + .40S ≤ 295    (pounds of hazelnuts)
                R ≥ 0, D ≥ 0, S ≥ 0
```

15) CF = number of cotton filled parkas to make, WF = wool filled, GF = goose-down filled
```
Maximize      24CF + 32WF + 36GF
subject to    50CF + 50WF + 50GF ≤ 40000  (sewing time - min)
               8CF +  9WF +  6GF ≤ 4800   (stuffing time - min)
                     12WF          ≤ 3600   (wool - oz)
                           12GF ≤ 2160   (goose-down - oz)
               CF ≥ 0, WF ≥ 0, GF ≥ 0
```

17) P1 = number of pill#1 to take per day, P2 = number of pill#2 to take per day

```
Minimize      0.20P1 + 0.35P2
subject to       7P1 +     3P2 ≥ 21     (iron)
                 2P1 +     2P2 ≥ 10     (potassium)
                 4P1 +    14P2 ≥ 40     (calcium)
                  P1 ≥ 0,  P2 ≥ 0
```

19) L = amount of low sulfur fuel to use each hour, H = amount of high sulfur fuel to use

```
Minimize      0.75L + 0.50H
subject to       3L +     6H ≤  9     (Sulfur dioxide)
                 3L +     3H ≥ 12     (kilowatts to produce)
                  L ≥ 0,  H ≥ 0
```

21) I = pounds of I, II = pounds of II, III = pounds of III

```
Minimize       .45I + 1.07II + .60III
subject to       I +    II +   III =    1     (make 1 pound)
               .12I + .26II + .10III ≥ .14    (crude protein)
               .02I + .31II          ≥ .08    (crude fat)
               .11I + .01II + .18III ≤ .055   (crude fiber)
               .20I + .08II + .06III ≤ .12    (moisture)
                 I ≥ 0,  II ≥ 0, III ≥ 0
```

23) A = amount of alloy A to use and let B,C,D,E be similarly defined

```
Minimize   3.9A + 4.1B +   6C + 6.03D + 7.7E
subject to  .1A +  .1B +  .4C +  .6D +  .3E = .3    (iron)
            .1A +  .3B +  .5C +  .3D +  .4E = .3    (copper)
            .8A +  .6B +  .1C +  .1D +  .3E = .4    (lead)
             A ≥ 0,  B ≥ 0,  C ≥ 0,  D ≥ 0,  E ≥ 0
```

END OF SECTION EXERCISES
LINEAR PROGRAMMING SECTION 4.2

1)   3)

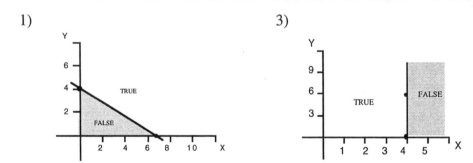

5) The maximum is 35 and occurs at (4,5). The minimum is 8 at (1,1).

7) He should serve 2 chocolate and 2 granola bars for a "smile" total of 18.

9) Serve 15 cups of punch and 25 cups of lemonade for a profit of $3.75.

11) Make 7 shawls and 3 afghans for a profit of $295.

13) Buy 5 regular and 4 large lockers for a storage capacity of 130 cubic feet.

15) Take 3 of pill 1 and 2 of pill 2 for a cost of $1.30.

17) NO SOLUTION

19) $30,000 in option A and $20,000 in option B.

1) Grant 2.1 million in car loans and 10.5 million in mortgages for a return of 1.0815 million.

3) Make on 1333.33 pounds of the regular mix for a profit of $5333.33.

5) Send 25 from warehouse 1 to store A, 75 from warehouse 2 to store A, 100 from warehouse 1 to store B and 100 from warehouse 2 to store C for a total cost of $34.50.

7) a) Melissa should invest $50,000 in stocks and $70,000 in bonds for a total of $10,900 in interest.

b. Melissa should NOT worry. 2% in either direction stays well within the bounds on the coefficients in the objective.

9) a) The company should make a weekly average of 148.8 wool-, 297.60 cotton- and 180 goose-filled jackets for a profit of $18,384.00.

b) The company should hire a stuffer.

c) Yes. Make 465 cotton- and 180 goose-filled jackets for a profit of $20,430.00.

d) No the answer won't change.

11) a) Make 100 easyrider and 20 economy chairs for a profit of $5,500.00.

b) An hour is worth $5 in wood working, $6 in finishing and $0 in upholstery. Add in the finishing shop.

c) The time in finishing does NOT become less valuable,.

d) We can lower the profit down to $14.71.

13) a). Make 40 vats of light, 10 dark and 30 ale for a profit of $380.00.

b) The profit for the premium must be raised to $14.01.

c) Make 25 vats of ale and 50 vats of dark for a profit of $325.00.

d) Increase the supply of malt because it has the highest shadow price.

15) a) Make an average 1333.33 of desk 1 and 66.67 of desk 4 per 6 month period for a profit of $18,666.67.

b). Add more hours in carpentry.

c) Make the profit on desk 2 $26.68.

d) We need to resolve: Make 800 of desk 3 and 40 of desk 4 for a profit of $16,000.

1) $x = 4, y = 0$, and Z(obj) = 20   3) $x = .3333, y = 3, z = 0$ and Z(obj) = 10.33

5) Build the factory and warehouse both in Raleigh for a profitability of $14 million.

7) Make 127 cotton, 300 wool and 180 goose down jackets for a profit of $19,128.

9) Make 4250 Kens and 4500 Skippers for a profit of $54,400.

11) Make 72.5 vats of pale ale for a profit of $7,250.

13) Fly sequences 3, 4, and 11 for a minimum cost of $18.